Nobel Physicists

ISBN 978-3-939201-10-6

The ongoing "NOBELS – Nobel Laureates photographed by Peter Badge" project is kindly supported by the Klaus Tschira Foundation.

"Nobel Physicists" has been published on the occasion of the 69th Lindau Nobel Laureate Meeting (30 June – 5 July 2019).

© 2019 Foundation Lindau Nobel Laureate Meetings

© Texts by the authors, photographs by Peter Badge

The biographical information has been taken from the Nobelprize.org webpage, with kind permission from The Nobel Foundation.

Lithography: artificial image, Berlin

Design: Studio @z23, Berlin

Printing: PieReg Druckcenter Berlin GmbH

Printed in Germany

Stiftung Lindauer Nobelpreisträgertagungen

Foundation Lindau Nobel Laureate Meetings

Legal domicile: 78465 Insel Mainau, Germany

Office: Lennart-Bernadotte-Haus, Alfred-Nobel-Platz 1, 88131 Lindau im Bodensee, Germany

foundation@lindau-nobel.org, www.lindau-nobel.org

Peter Badge · Nobel Physicists

Foreword by Countess Bettina Bernadotte
Preface by Angela Merkel
Introduction by Brian Schmidt
Essay by Nikolaus Turner

Honouring and Encouraging Excellence – The Lindau Nobel Laureate Meetings

The portraits displayed in this book form part of a long term project supported by the Foundation Lindau Nobel Laureate Meetings. The portrayed lady and gentlemen deserve our recognition and admiration for their constant striving for excellence. But, as film director and photographer Wim Wenders put it in his epilogue for the book *Nobel Heroes,* publoished by Steidl, "you don't get the Nobel Prize just for excellence in your fields, you get it for surpassing yourself, for hardship endured, for service beyond the call of duty." To encourage younger generations of researchers to strive for excellence and maintain their passion for science is one of the core aims of the Lindau Nobel Laureate Meetings, where most of the portraits were taken. With a history dating back to 1951, the Lindau Meetings have been bringing together the most esteemed scientists of our times with brilliant young scientists and researchers from all over the world. In 2019, the 69th Lindau Nobel Laureate Meeting will focus on Physics. The range of lecture and discussion topics, gravitational waves, encompassing the dark side of the universe, laser physics and graphene as part of materials physics, promises to spark inspiring debates and further reflections among all participants.

Countess Bettina Bernadotte af Wisborg
President of the Council for the Lindau Nobel Laureate Meetings

Preface

The Lindau Nobel Laureate Meeting is far more than a nice tradition. For nearly 70 years it has offered excellent young scientists a unique opportunity to discuss current research issues with world-renowned luminaries in their field and to cultivate international contacts. The personal encounters with Nobel Laureates are a source of inspiration and encouragement to promising young scientists from all over the world. Their eyes are opened to new approaches to their own scientific work. It is an investment in the future in the best sense, because progress in science and research is now and will remain the key to competitiveness and prosperity. The Lindau Nobel Laureate Meeting is thus an excellent showcase for science in Germany.

The focus of this year's meeting is on physics. Nearly 40 Nobel Laureates will come together in Lindau with around 580 young scientists from almost 90 countries. This unique meeting of generations of scientists was the occasion for photographer Peter Badge to produce an opulent volume of photographs with impressive portraits of Nobel Laureates in Physics. Their revolutionary discoveries, insights and inventions have become part of our daily lives in many ways. Without their work, the world today would not be the same.

This volume of photos is a successful homage to renowned scientific figures. Let it be an inspiration to young scientists to further outstanding achievements in research.

Angela Merkel
Chancellor of the Federal Republic of Germany

Introduction

Education is the foundation of prosperity. It gives citizens the power of self-determination, it enables nations to provide the goods and services which we depend on in our daily lives, and it gives the world the new ideas that continually revolutionise the limits of what we can do. A huge number of these ideas come from scientists, and so while helping ensure we have a supply of highly capable scientists is not the only reason for a high standard of education, it is an incredibly important outcome. Certainly education has been paramount in my life.

I grew up in the mountains of Montana where I received a wonderfully uncomplicated and empowering primary education. I learned the basics of English, history, science, and math in an environment that I remember as both being fun and stimulating. Like most things in life, school is most effective when it is enjoyable. High School, to this day, remains one of the best times of my life. My school provided a sophisticated environment to learn not just my school subjects, but how to take on all the world has to offer. It provided me the confidence to undertake a degree in Astronomy, knowing that I probably would not become an astronomer, but that the education I gained would open up a universe of possibilities for me in my life.

Five years later I found myself in the Alps at a school dedicated to supernovae. The mixture of young graduate students and many of the most eminent astronomers in the world led to what I consider to be the most important weeks of my academic life. Suddenly I felt transformed into a real scientist. I was full of new pieces of knowledge, exciting ideas, and I left overflowing with optimism. The relationships I set up during this school were the basis of what lead to our discovery of the accelerating universe, and remain the foundation of much of the work I continue to do to this day.

Every year the Lindau Meetings bring together Nobel Laureates and young scientists from around the world for a week of lectures, discussions, and socialising in the Alps. This is a unique educational event that reaches out to the pinnacle of our young scientists. From this grouping will be many of the future scientific leaders of the world, and in all likelihood a Nobel Laureate or two.

2019 will already be my sixth visit to Lindau, and it will be part of another experience in the Alps like that I remember so fondly from nearly 30 years ago. And since education is a lifelong process, I will again learn and benefit from this school every bit as much as the students.

An important part of the Lindau Meetings is for students to see that Nobel Laureates are real people, just like everyone else, and that we are no different from them. Humanising science is one of the most important ways of ensuring its future relevance, by making it both attractive to budding scientists, and accessible to the general public.

Another project of the Lindau Nobel Laureate Meetings to humanise science is its programme to take photographic portraits of all of the Nobel Laureates. It was a cathartic experience for me walking around Mount Stromlo with Peter Badge telling him stories of my life while he captured the moment on film. The portraits capture the people behind the Nobel Prizes and give us a glimpse of the excitement of their discoveries. I hope you will share with me the fascination of looking through the history of physics as seen through the portraits of Nobel Laureates. This is a wonderful way that the Foundation helps us all share in the excitement that is scientific discovery.

Brian Schmidt
Nobel Laureate in Physics 2011

Alexei A. Abrikosov *1928 †2017

The Nobel Prize in Physics 2003 was awarded jointly to Alexei A. Abrikosov, Vitaly L. Ginzburg and Anthony J. Leggett "for pioneering contributions to the theory of superconductors and superfluids."

Isamu Akasaki *1929

The Nobel Prize in Physics 2014 was awarded jointly to Isamu Akasaki, Hiroshi Amano and Shuji Nakamura "for the invention of efficient blue light-emitting diodes which has enabled bright and energy-saving white light sources."

Zhores I. Alferov *1930 †2019

The Nobel Prize in Physics 2000 was awarded "for basic work on information and communication technology" with one half jointly to Zhores I. Alferov and Herbert Kroemer "for developing semiconductor heterostructures used in high-speed- and opto-electronics" and the other half to Jack S. Kilby "for his part in the invention of the integrated circuit."

Hiroshi Amano *1960

The Nobel Prize in Physics 2014 was awarded jointly to Isamu Akasaki, Hiroshi Amano and Shuji Nakamura "for the invention of efficient blue light-emitting diodes which has enabled bright and energy-saving white light sources."

Philip Warren Anderson *1923

The Nobel Prize in Physics 1977 was awarded jointly to Philip Warren Anderson, Sir Nevill Francis Mott and John Hasbrouck van Vleck "for their fundamental theoretical investigations of the electronic structure of magnetic and disordered systems."

Arthur Ashkin *1922

The Nobel Prize in Physics 2018 was awarded "for groundbreaking inventions in the field of laser physics" with one half to Arthur Ashkin "for the optical tweezers and their application to biological systems", the other half jointly to Gérard Mourou and Donna Strickland "for their method of generating high-intensity, ultra-short optical pulses."

Barry C. Barish *1936

The Nobel Prize in Physics 2017 was divided, one half awarded to Rainer Weiss, the other half jointly to Barry C. Barish and Kip S. Thorne "for decisive contributions to the LIGO detector and the observation of gravitational waves."

J. Georg Bednorz *1950

The Nobel Prize in Physics 1987 was awarded jointly to J. Georg Bednorz and K. Alexander Müller "for their important break-through in the discovery of superconductivity in ceramic materials."

Hans Albrecht Bethe *1906 †2005

The Nobel Prize in Physics 1967 was awarded to Hans Bethe "for his contributions to the theory of nuclear reactions, especially his discoveries concerning the energy production in stars."

Gerd Binnig *1947

The Nobel Prize in Physics 1986 was divided, one half awarded to Ernst Ruska "for his fundamental work in electron optics, and for the design of the first electron microscope", the other half jointly to Gerd Binnig and Heinrich Rohrer "for their design of the scanning tunneling microscope."

Nicolaas Bloembergen *1920 †2017

The Nobel Prize in Physics 1981 was divided, one half jointly to Nicolaas Bloembergen and Arthur Leonard Schawlow "for their contribution to the development of laser spectroscopy" and the other half to Kai M. Siegbahn "for his contribution to the development of high-resolution electron spectroscopy."

Aage Niels Bohr *1922 †2009

The Nobel Prize in Physics 1975 was awarded jointly to Aage Niels Bohr, Ben Roy Mottelson and Leo James Rainwater "for the discovery of the connection between collective motion and particle motion in atomic nuclei and the development of the theory of the structure of the atomic nucleus based on this connection."

Willard S. Boyle *1924 †2011

The Nobel Prize in Physics 2009 was divided, one half awarded to Charles Kuen Kao "for groundbreaking achievements concerning the transmission of light in fibers for optical communication", the other half jointly to Willard S. Boyle and George E. Smith "for the invention of an imaging semiconductor circuit – the CCD sensor."

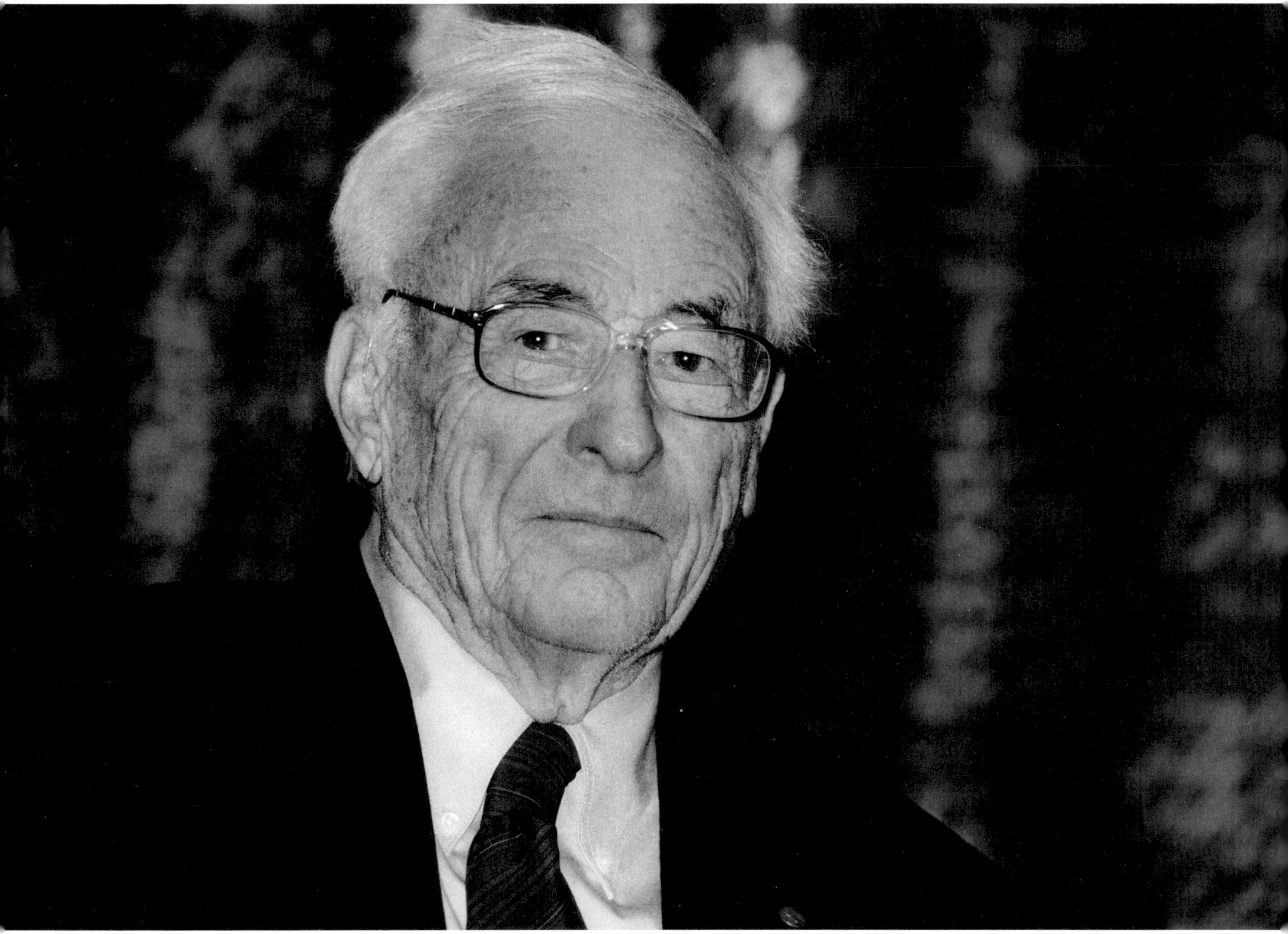

Owen Chamberlain *1920   †2006

The Nobel Prize in Physics 1959 was awarded jointly to Emilio Gino Segrè and Owen Chamberlain "for their discovery of the antiproton."

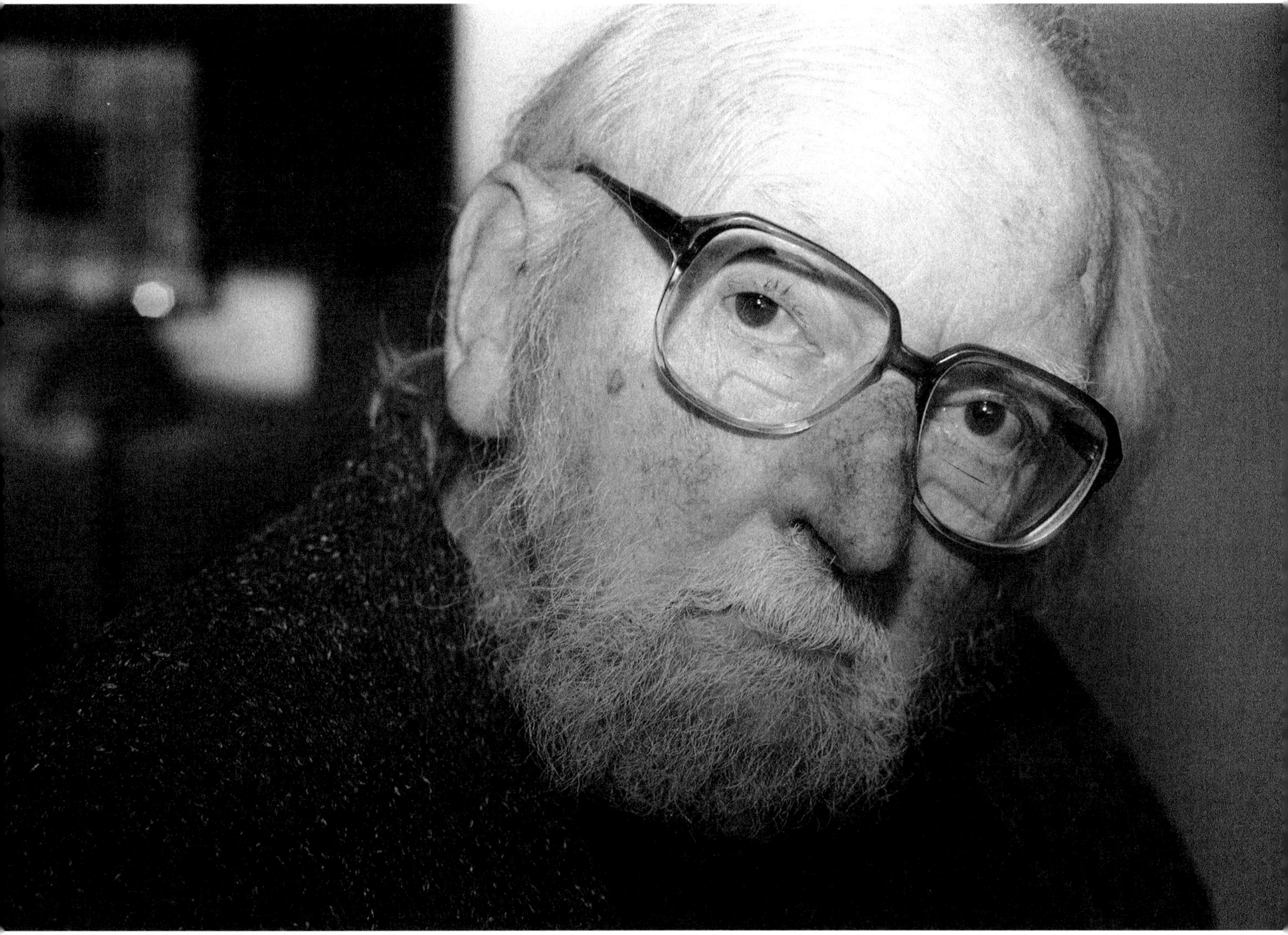

Georges Charpak *1924   †2010

The Nobel Prize in Physics 1992 was awarded to Georges Charpak "for his invention and development of particle detectors, in particular the multiwire proportional chamber."

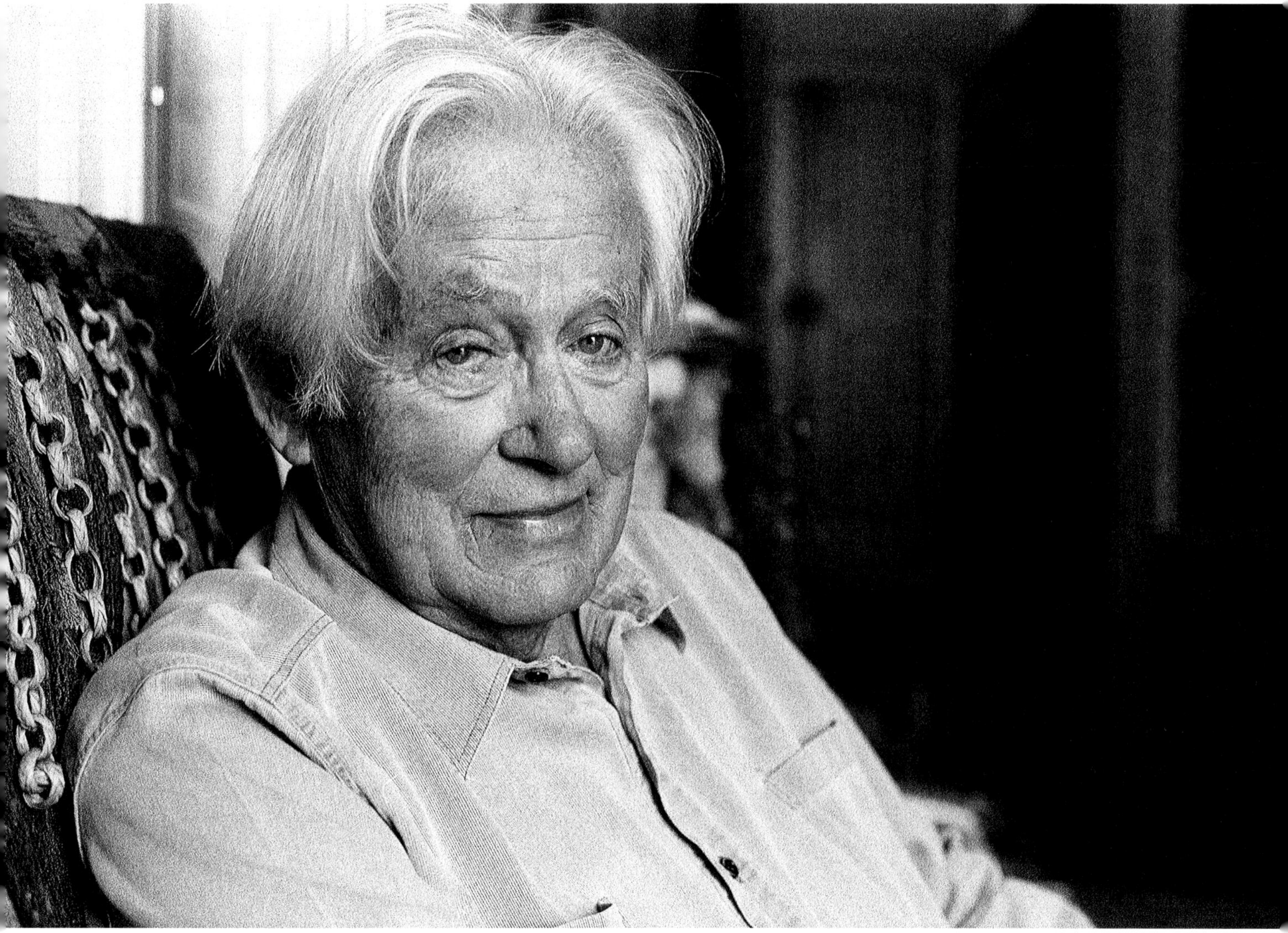

Steven Chu *1948

The Nobel Prize in Physics 1997 was awarded jointly to Steven Chu, Claude Cohen-Tannoudji and William D. Phillips "for development of methods to cool and trap atoms with laser light."

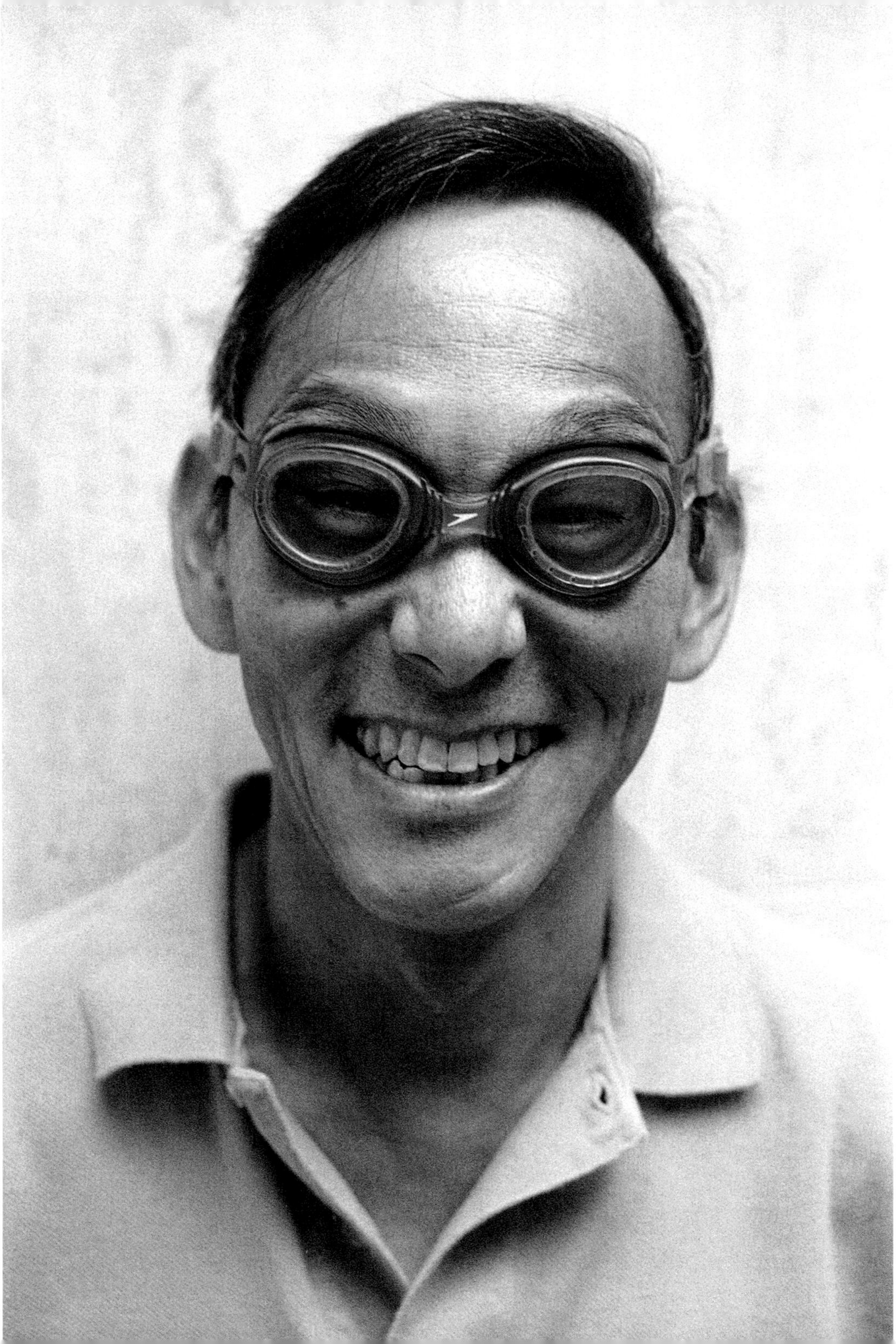

Claude Cohen-Tannoudji *1933

The Nobel Prize in Physics 1997 was awarded jointly to Steven Chu, Claude Cohen-Tannoudji and William D. Phillips "for development of methods to cool and trap atoms with laser light."

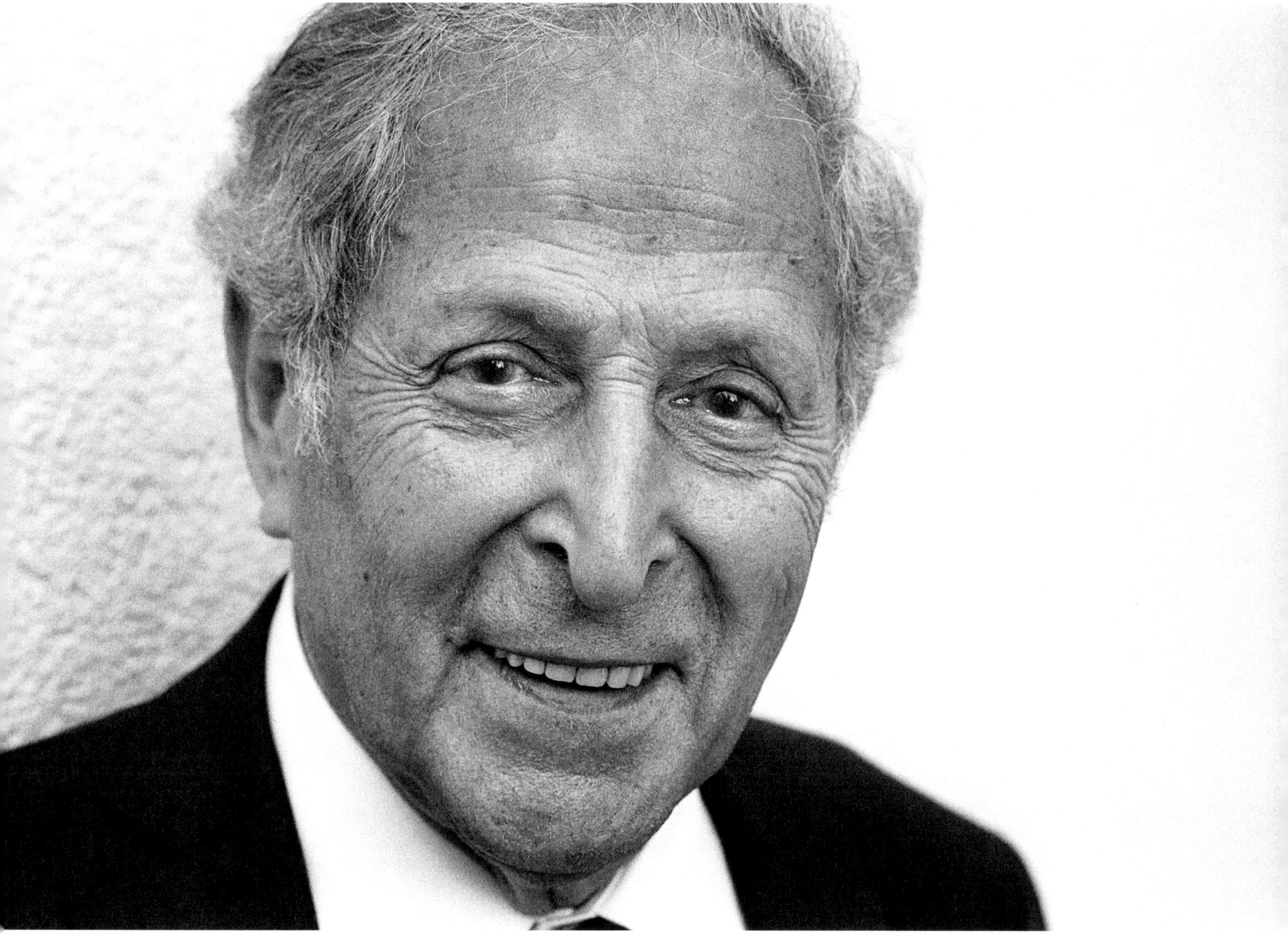

Leon Neil Cooper *1930

The Nobel Prize in Physics 1972 was awarded jointly to John Bardeen, Leon Neil Cooper and John Robert Schrieffer "for their jointly developed theory of superconductivity, usually called the BCS-theory."

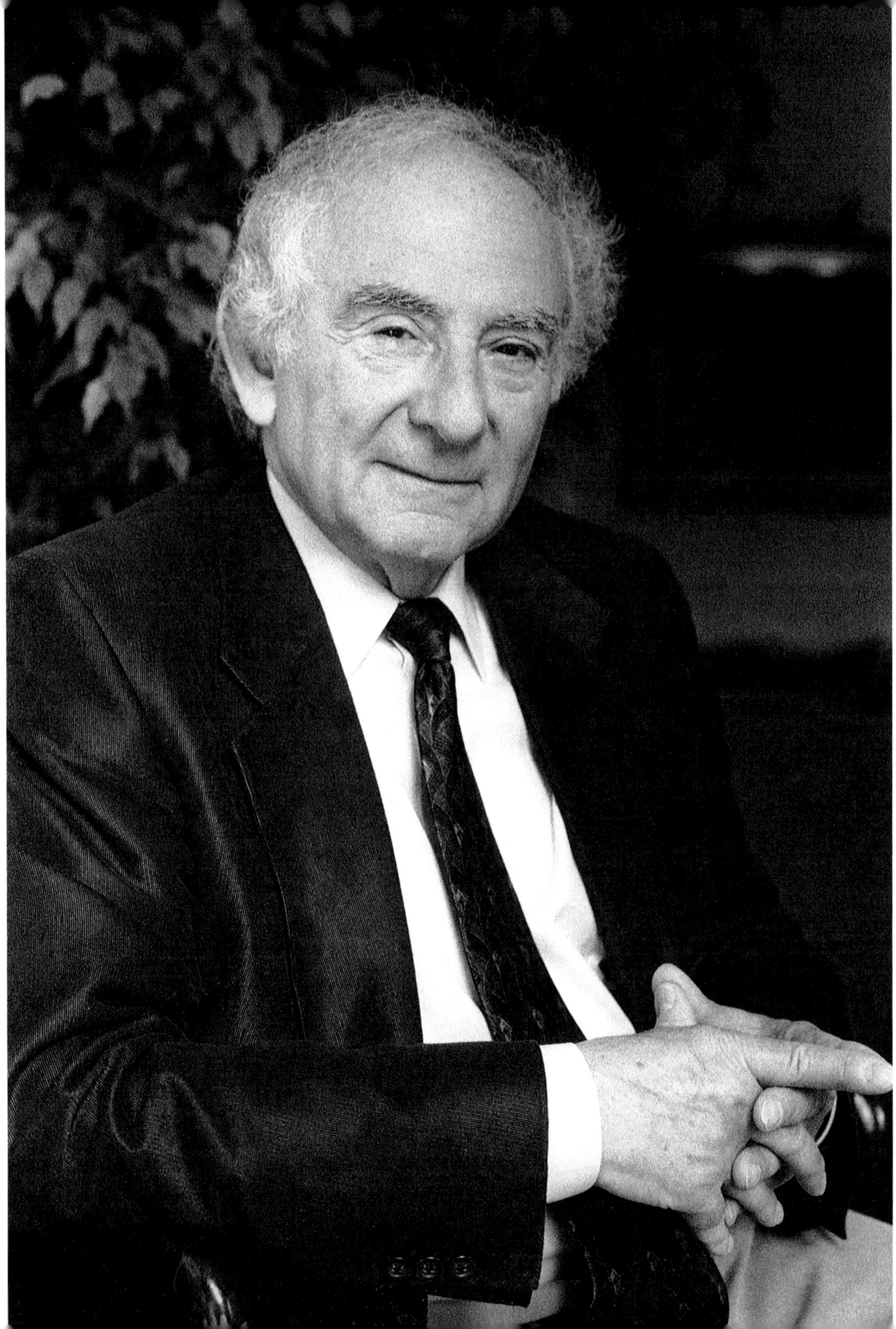

Eric A. Cornell *1961

The Nobel Prize in Physics 2001 was awarded jointly to Eric A. Cornell, Wolfgang Ketterle and Carl E. Wieman "for the achievement of Bose-Einstein condensation in dilute gases of alkali atoms, and for early fundamental studies of the properties of the condensates."

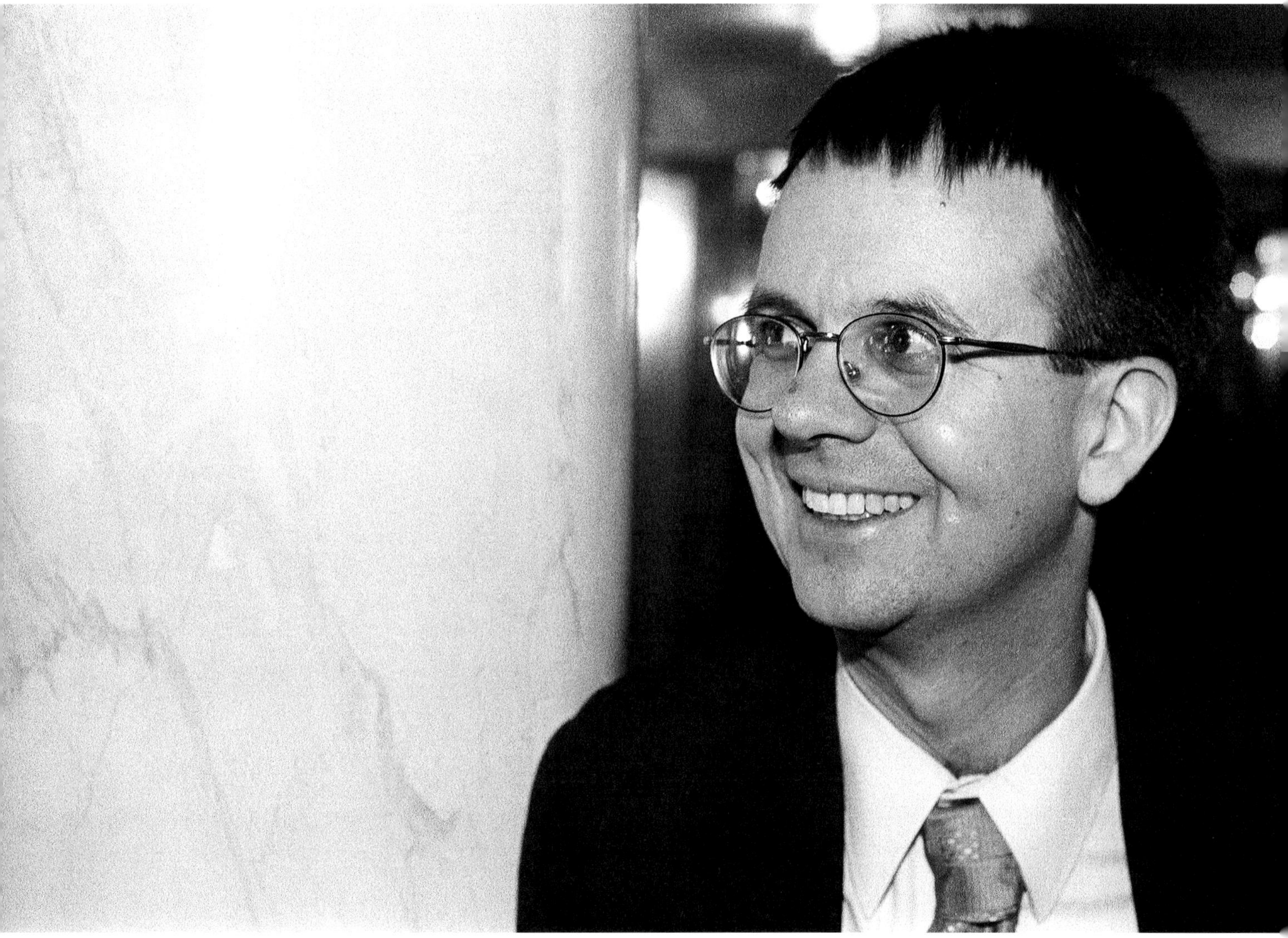

James Watson Cronin *1931 †2016

The Nobel Prize in Physics 1980 was awarded jointly to James Watson Cronin and Val Logsdon Fitch "for the discovery of violations of fundamental symmetry principles in the decay of neutral K-mesons."

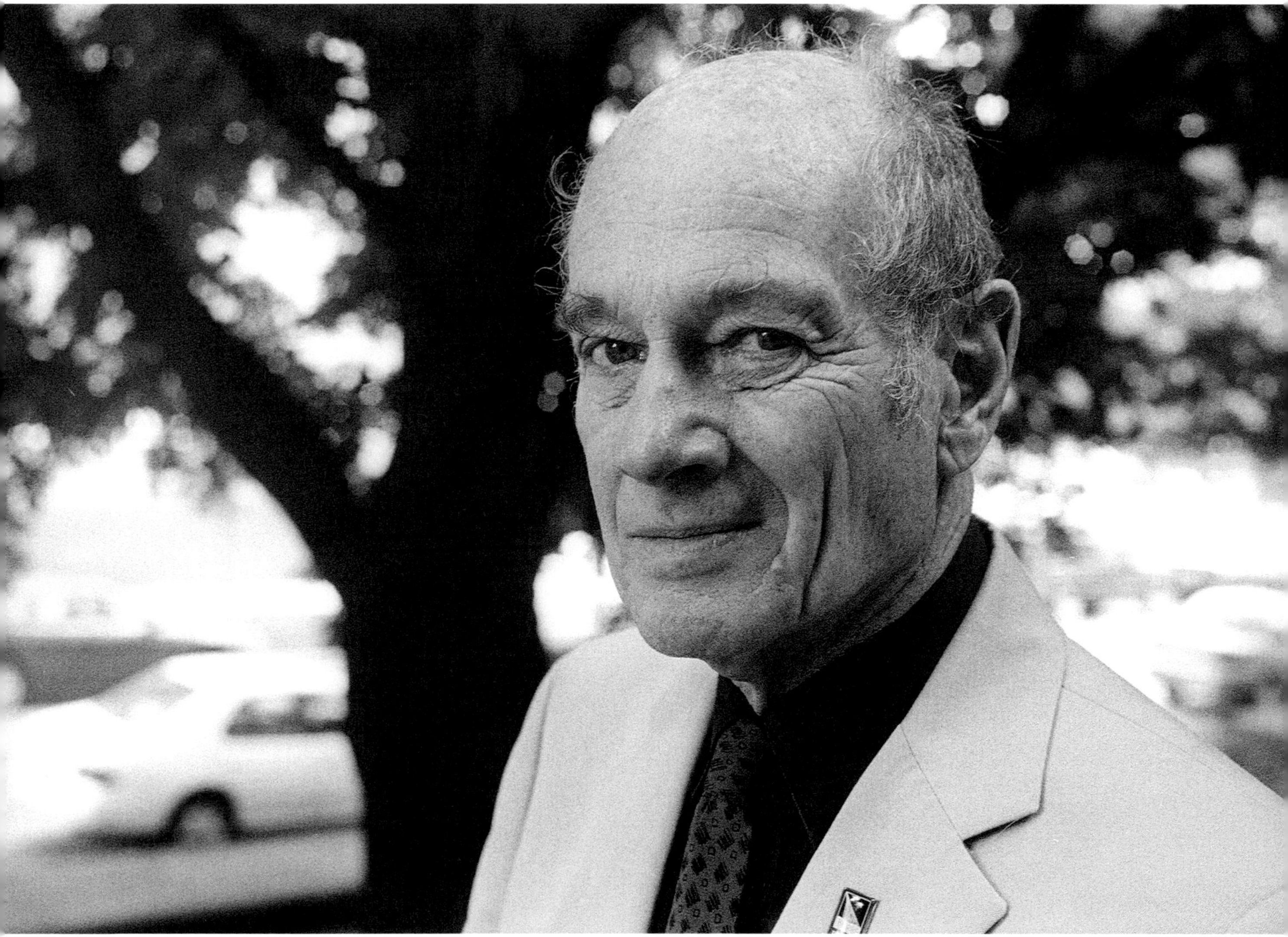

Raymond Davis Jr. *1914 †2006

The Nobel Prize in Physics 2002 was divided, one half jointly to Raymond Davis Jr. and Masatoshi Koshiba "for pioneering contributions to astrophysics, in particular for the detection of cosmic neutrinos" and the other half to Riccardo Giacconi "for pioneering contributions to astrophysics, which have led to the discovery of cosmic X-ray sources."

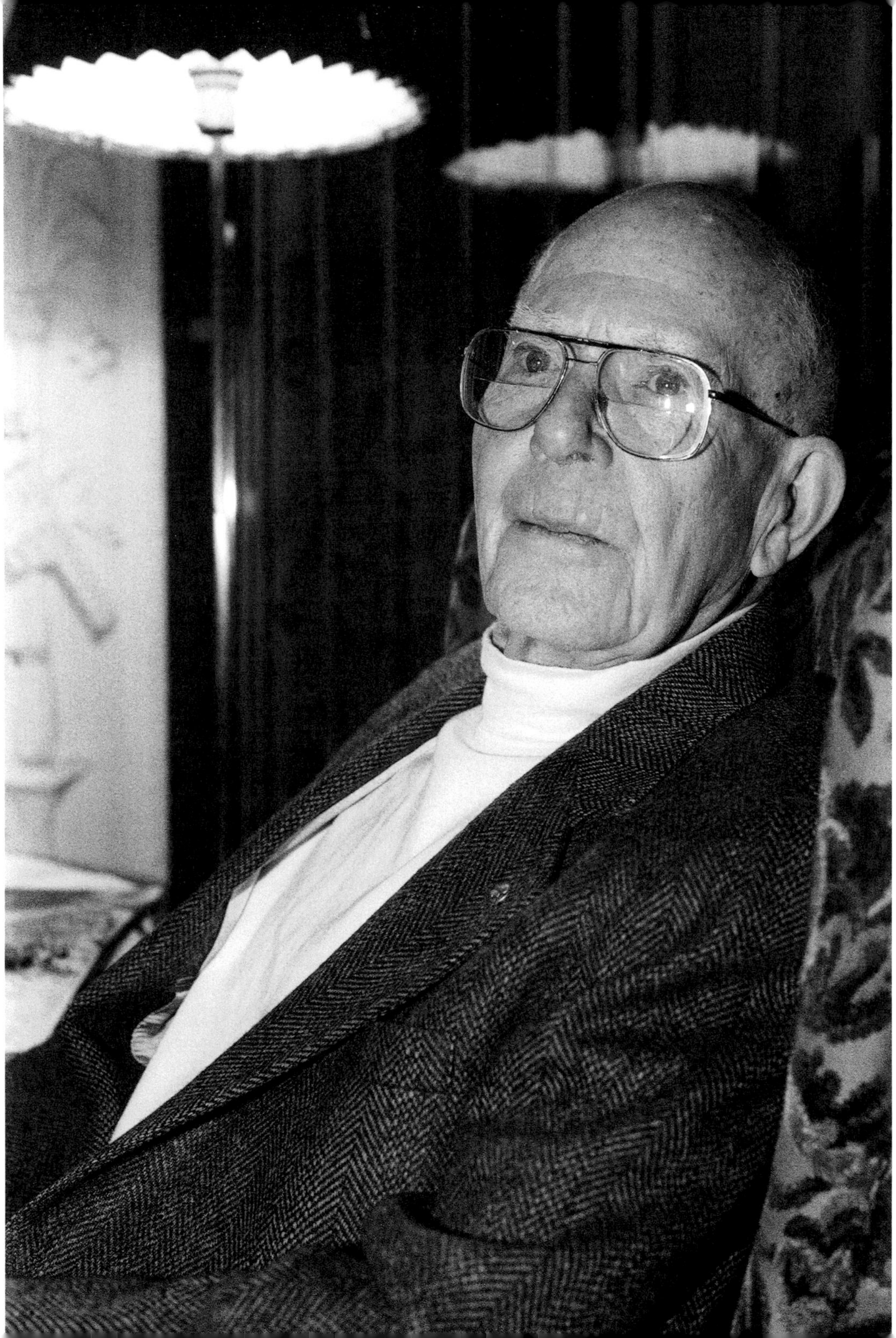

Hans G. Dehmelt *1922 †2017

The Nobel Prize in Physics 1989 was divided, one half awarded to Norman F. Ramsey "for the invention of the separated oscillatory fields method and its use in the hydrogen maser and other atomic clocks", the other half jointly to Hans G. Dehmelt and Wolfgang Paul "for the development of the ion trap technique."

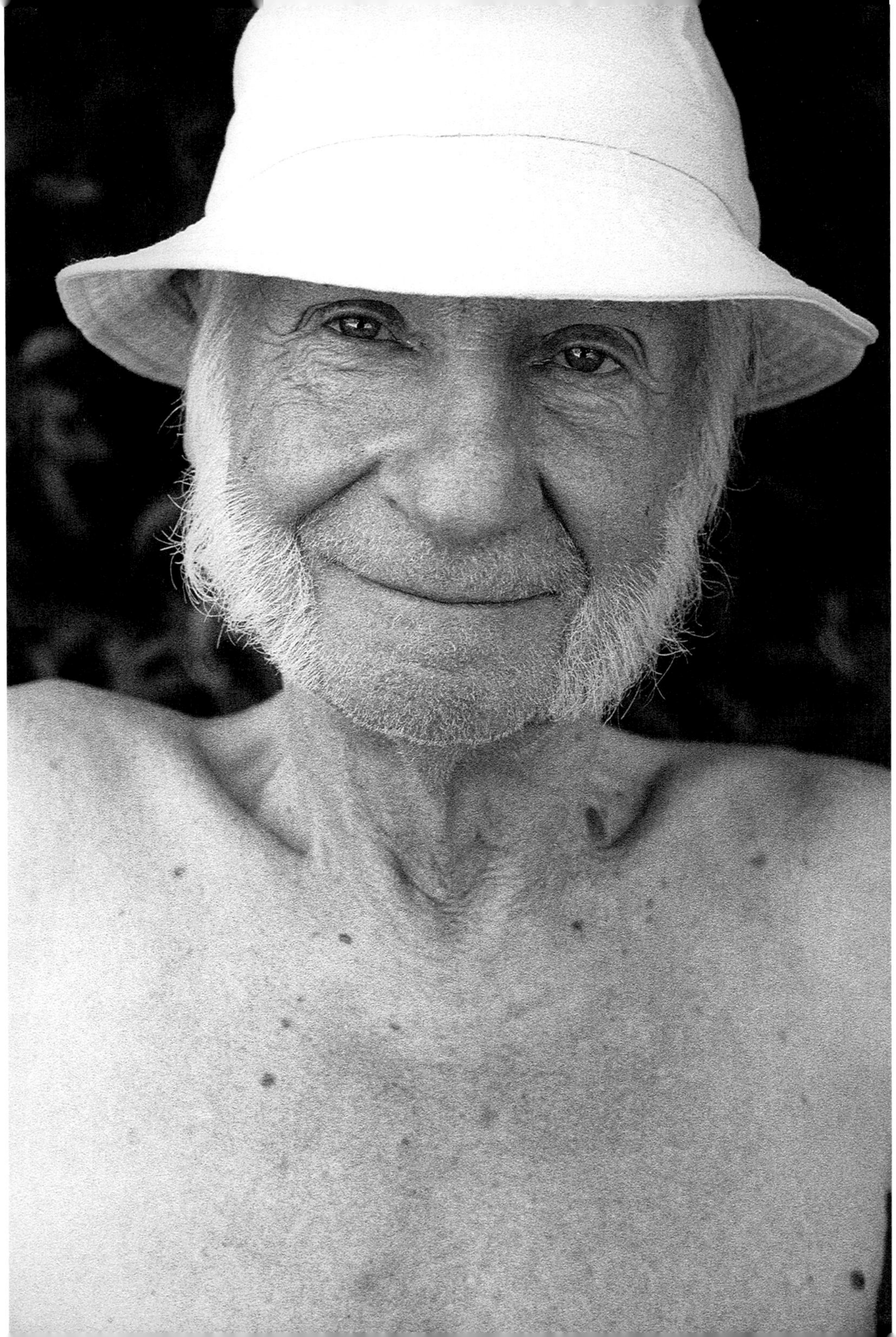

François Englert *1932

The Nobel Prize in Physics 2013 was awarded jointly to François Englert and Peter W. Higgs "for the theoretical discovery of a mechanism that contributes to our understanding of the origin of mass of subatomic particles, and which recently was confirmed through the discovery of the predicted fundamental particle, by the ATLAS and CMS experiments at CERN's Large Hadron Collider."

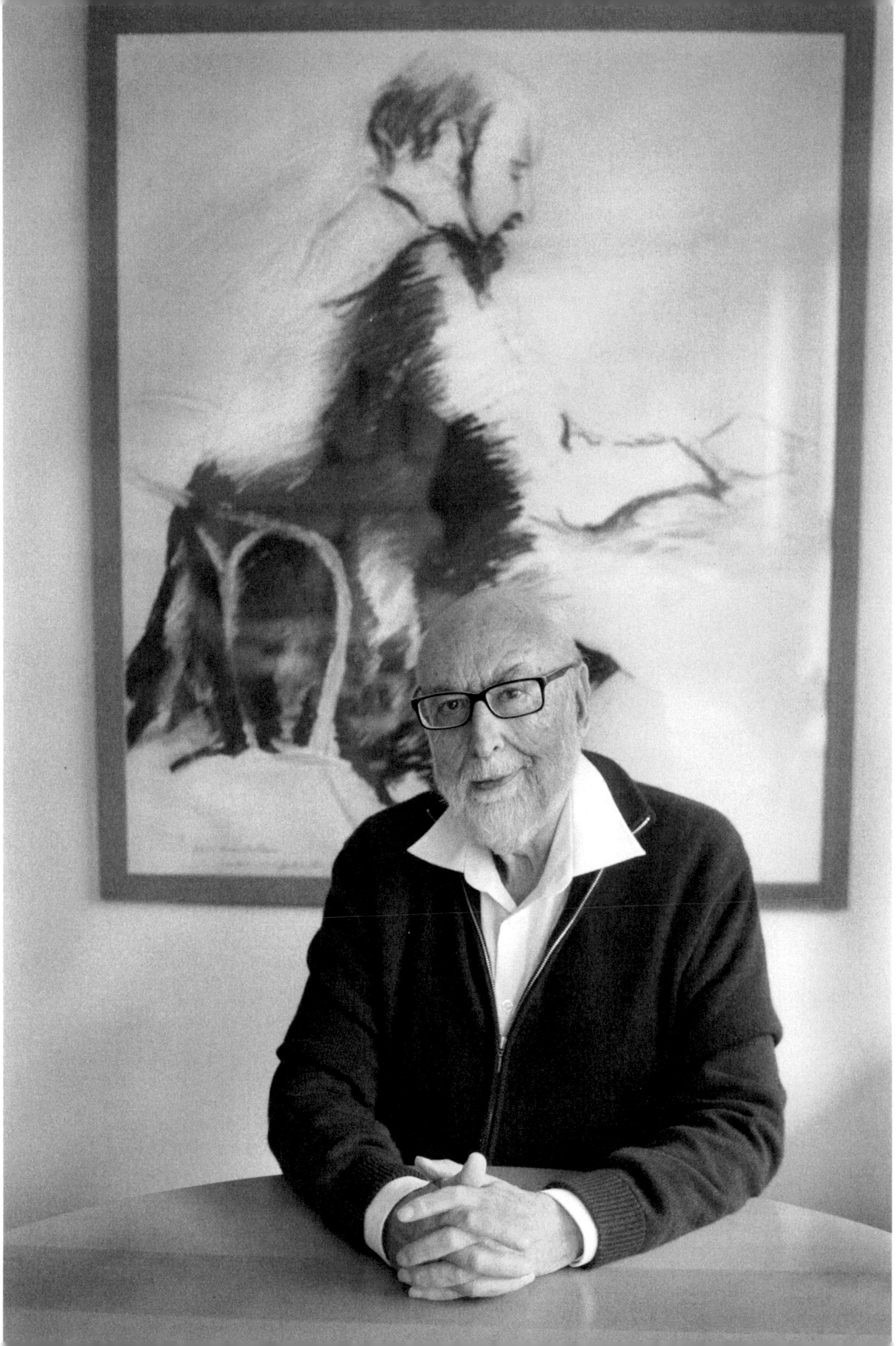

Leo Esaki *1925

The Nobel Prize in Physics 1973 was divided, one half jointly to Leo Esaki and Ivar Giaever "for their experimental discoveries regarding tunneling phenomena in semiconductors and superconductors, respectively" and the other half to Brian David Josephson "for his theoretical predictions of the properties of a supercurrent through a tunnel barrier, in particular those phenomena which are generally known as the Josephson effects."

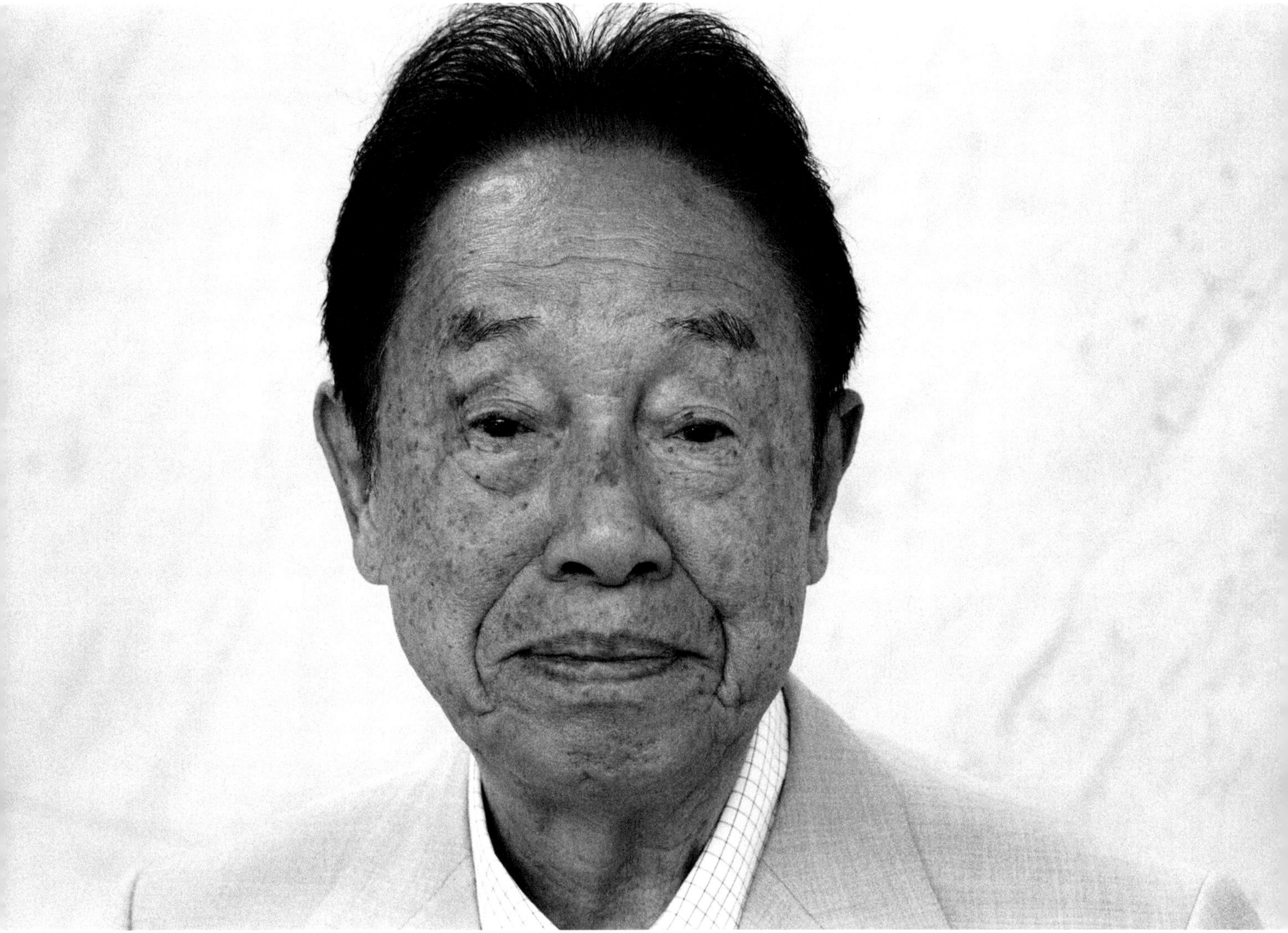

Albert Fert *1938

The Nobel Prize in Physics 2007 was awarded jointly to Albert Fert and Peter Grünberg "for the discovery of Giant Magnetoresistance."

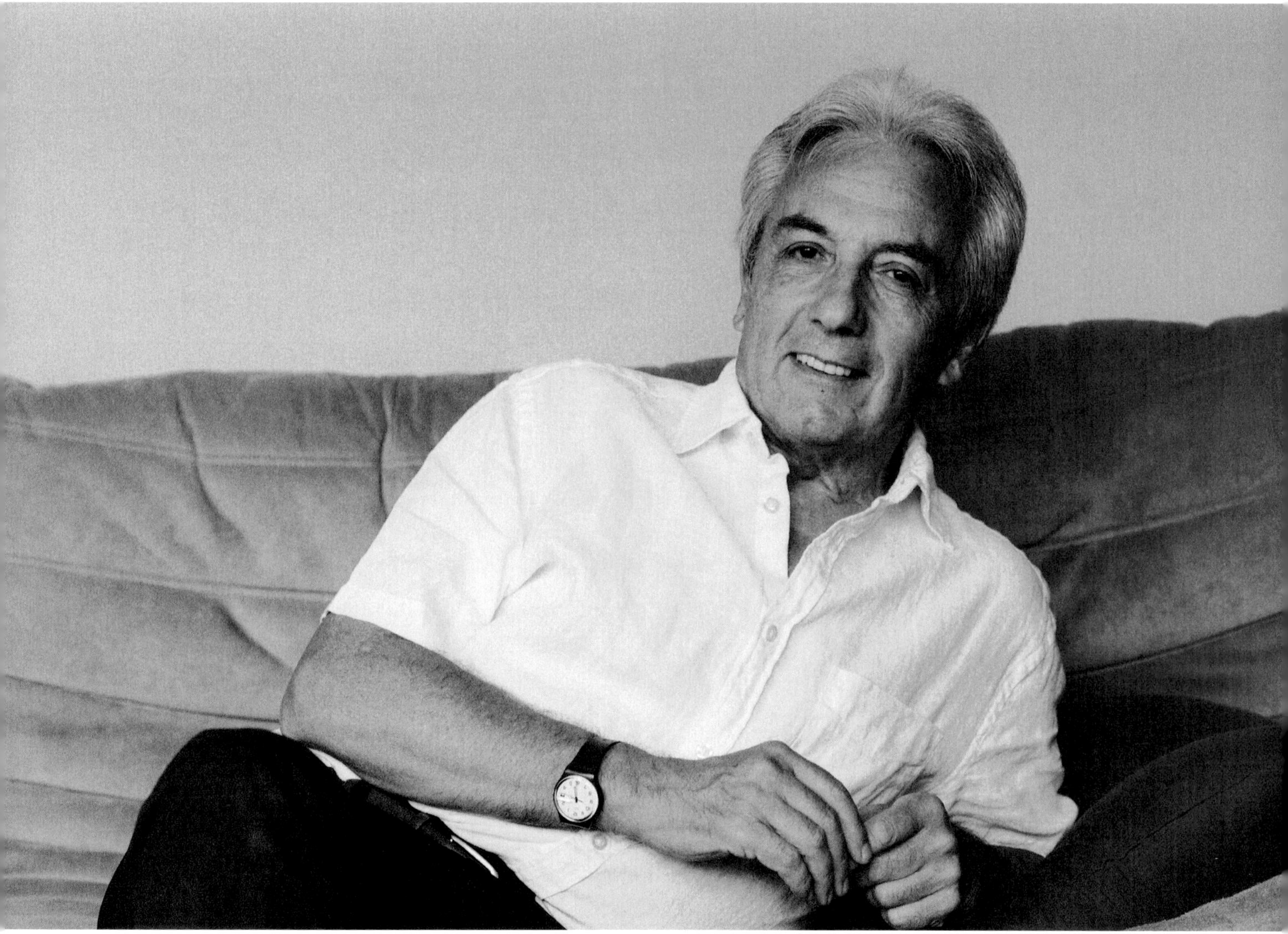

Val Logsdon Fitch *1923 †2015

The Nobel Prize in Physics 1980 was awarded jointly to James Watson Cronin and Val Logsdon Fitch "for the discovery of violations of fundamental symmetry principles in the decay of neutral K-mesons."

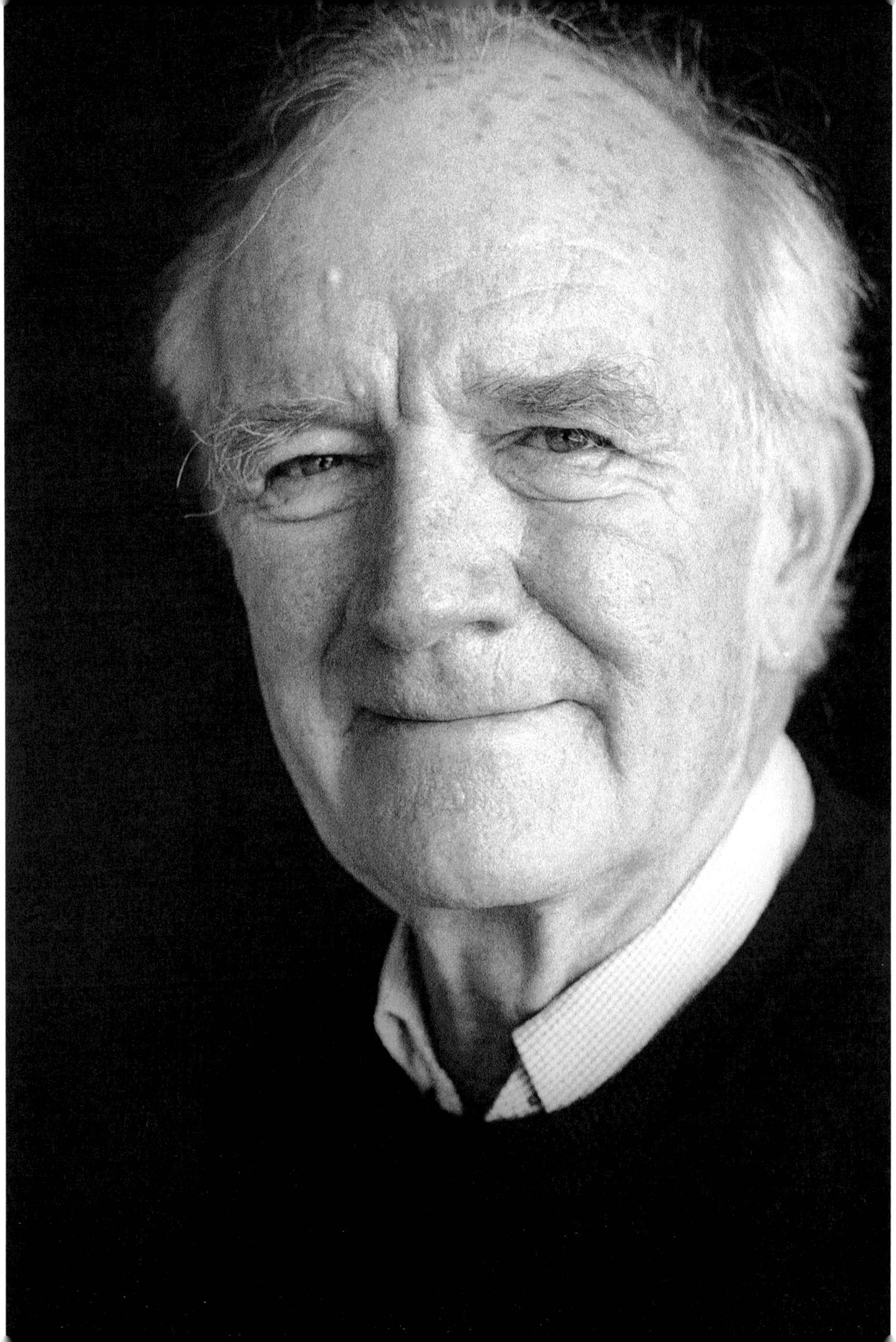

Jerome I. Friedman *1930

The Nobel Prize in Physics 1990 was awarded jointly to Jerome I. Friedman, Henry W. Kendall and Richard E. Taylor "for their pioneering investigations concerning deep inelastic scattering of electrons on protons and bound neutrons, which have been of essential importance for the development of the quark model in particle physics."

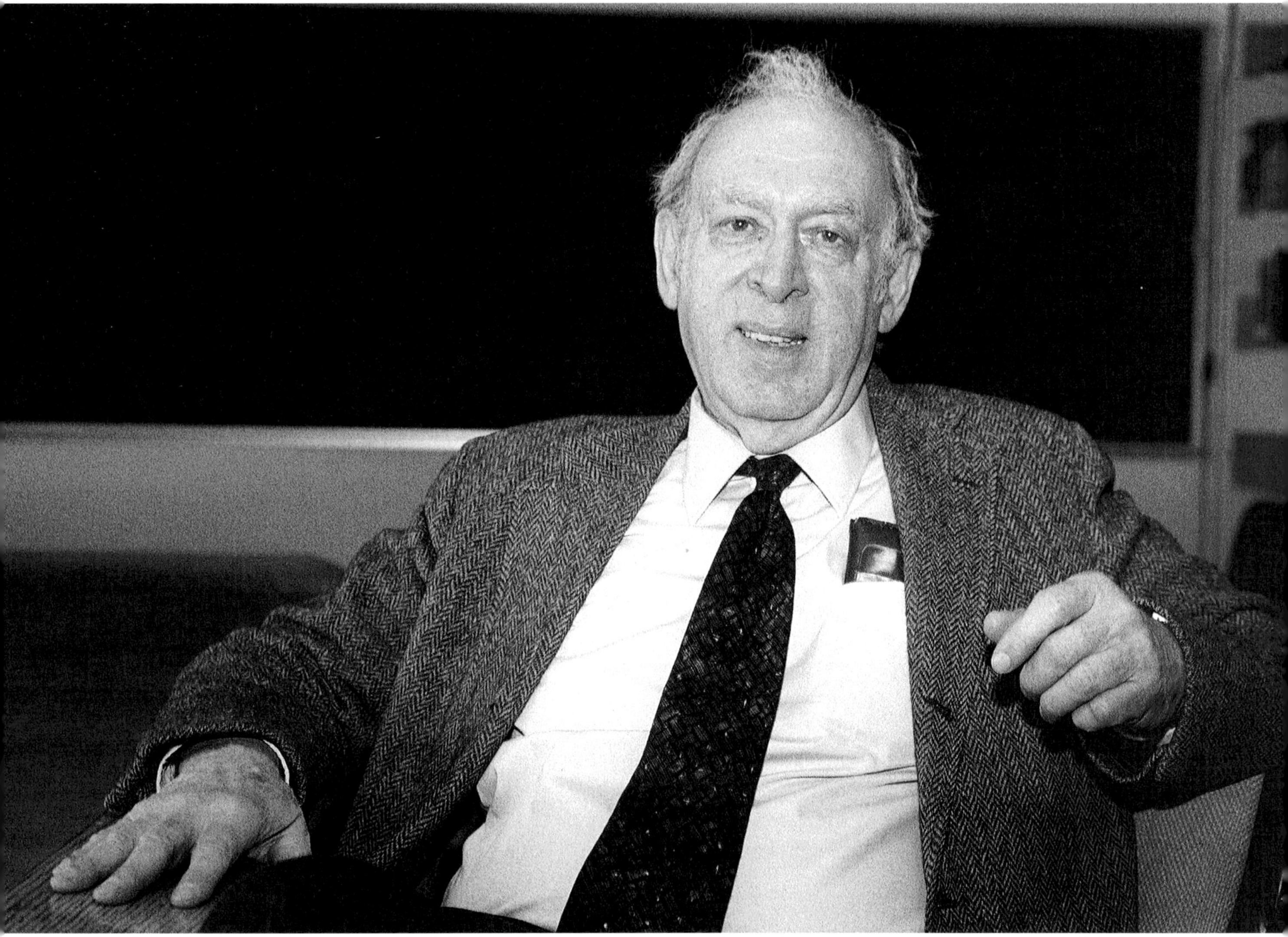

Andre Geim *1958

The Nobel Prize in Physics 2010 was awarded jointly to Andre Geim and Konstantin Novoselov "for groundbreaking experiments regarding the two-dimensional material graphene."

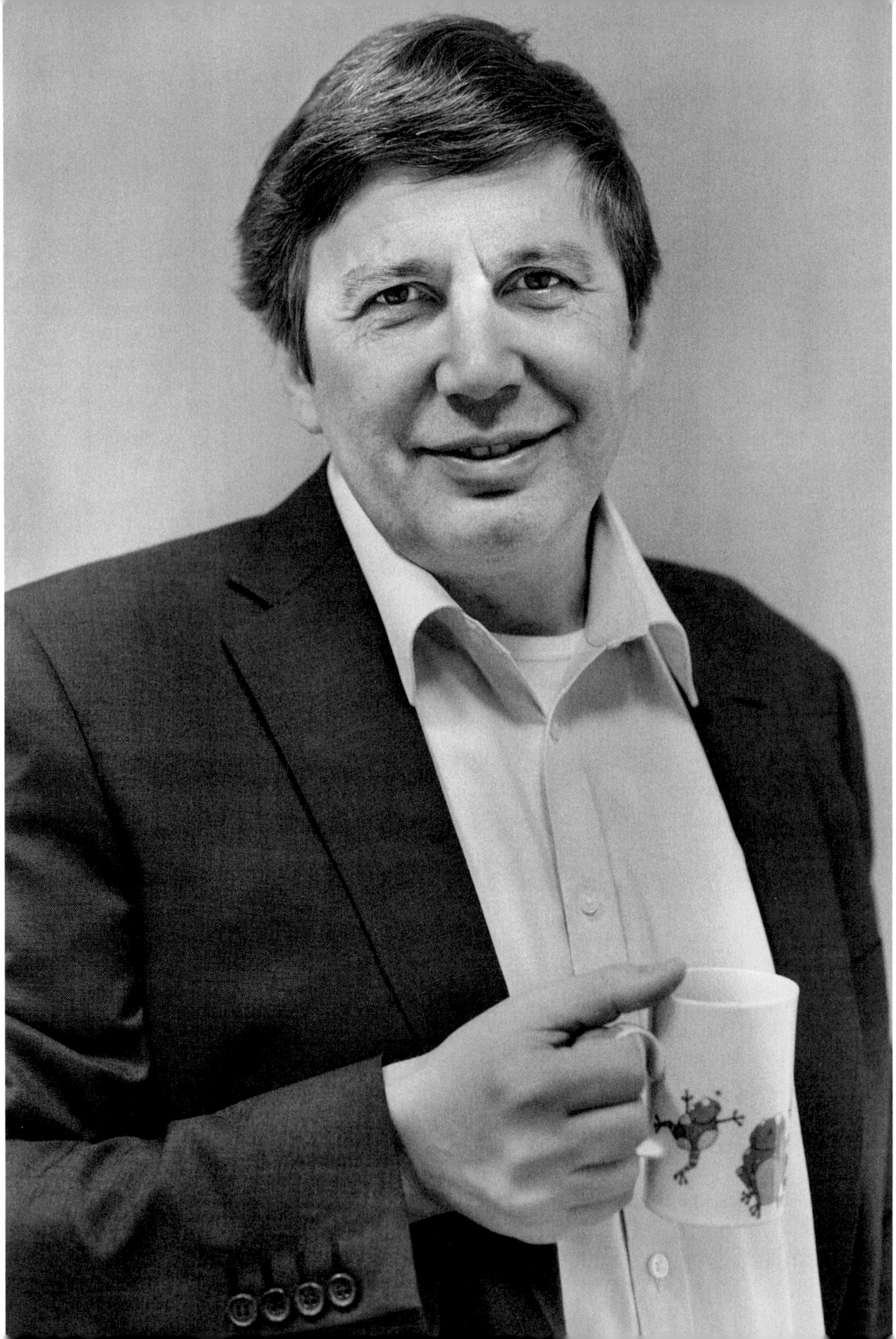

Murray Gell-Mann *1929 †2019

The Nobel Prize in Physics 1969 was awarded to Murray Gell-Mann "for his contributions and discoveries concerning the classification of elementary particles and their interactions."

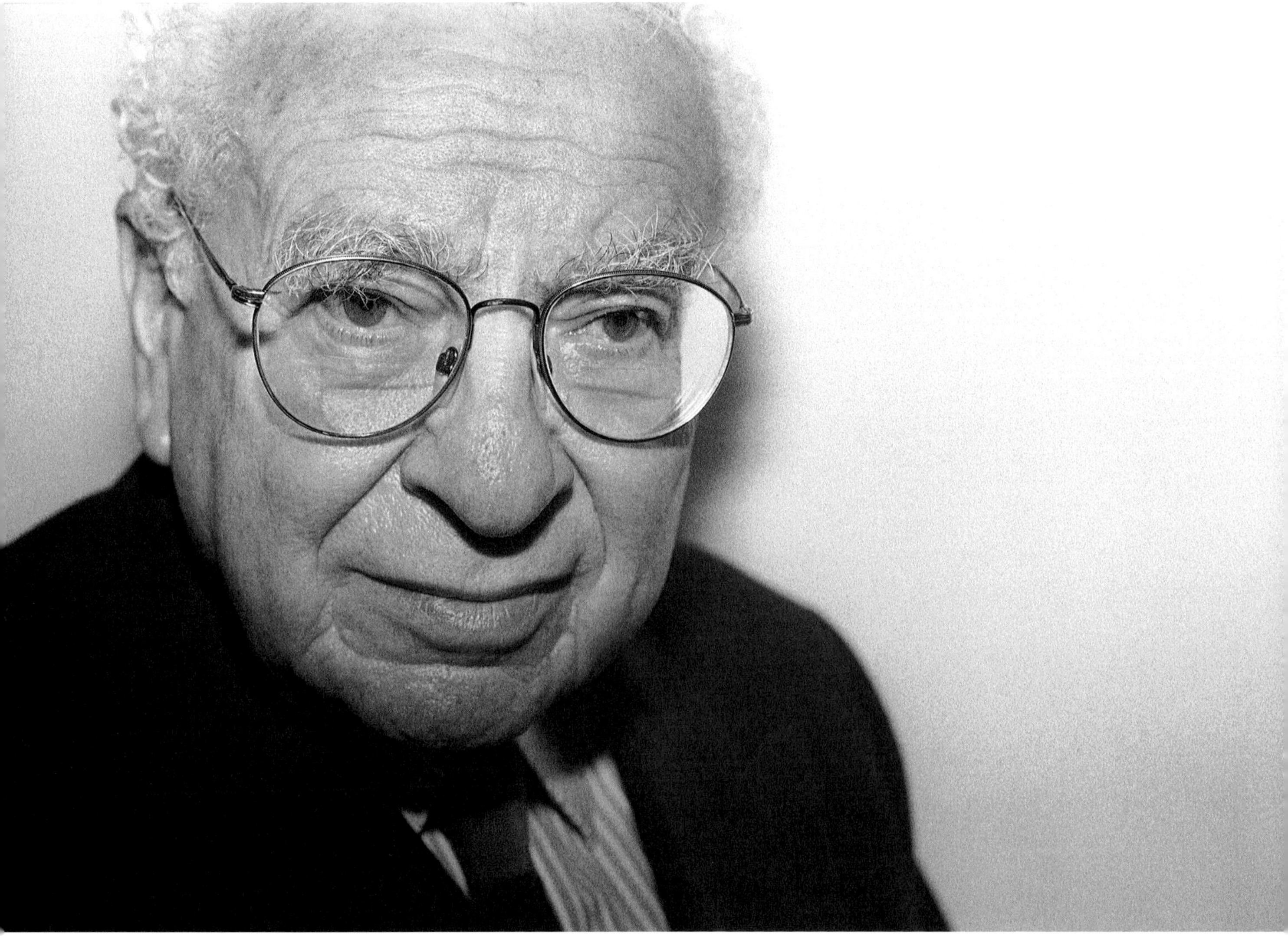

Pierre-Gilles de Gennes *1932 †2007

The Nobel Prize in Physics 1991 was awarded to Pierre-Gilles de Gennes "for discovering that methods developed for studying order phenomena in simple systems can be generalized to more complex forms of matter, in particular to liquid crystals and polymers."

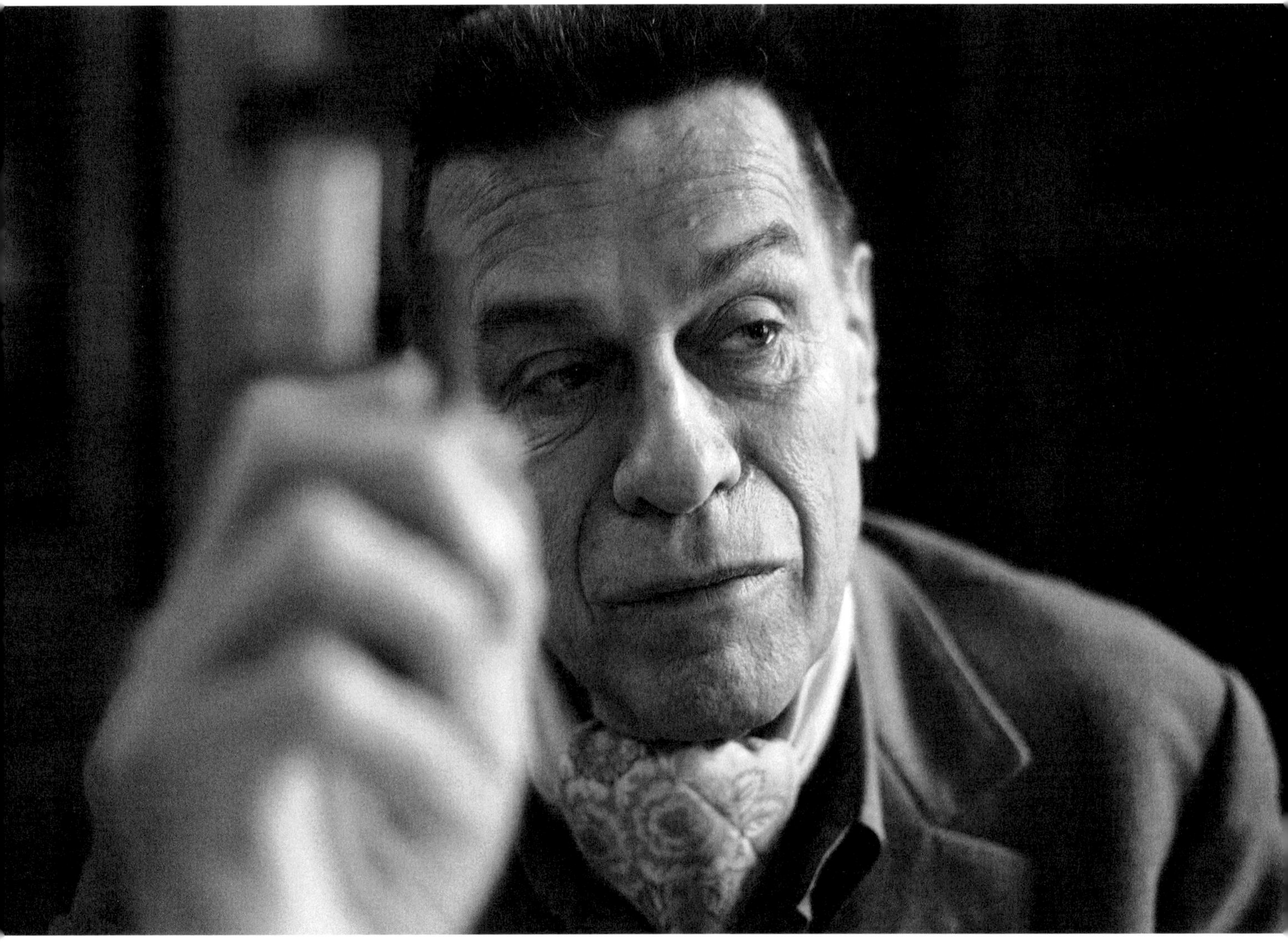

Riccardo Giacconi *1931 †2018

The Nobel Prize in Physics 2002 was divided, one half jointly to Raymond Davis Jr. and Masatoshi Koshiba "for pioneering contributions to astrophysics, in particular for the detection of cosmic neutrinos" and the other half to Riccardo Giacconi "for pioneering contributions to astrophysics, which have led to the discovery of cosmic X-ray sources."

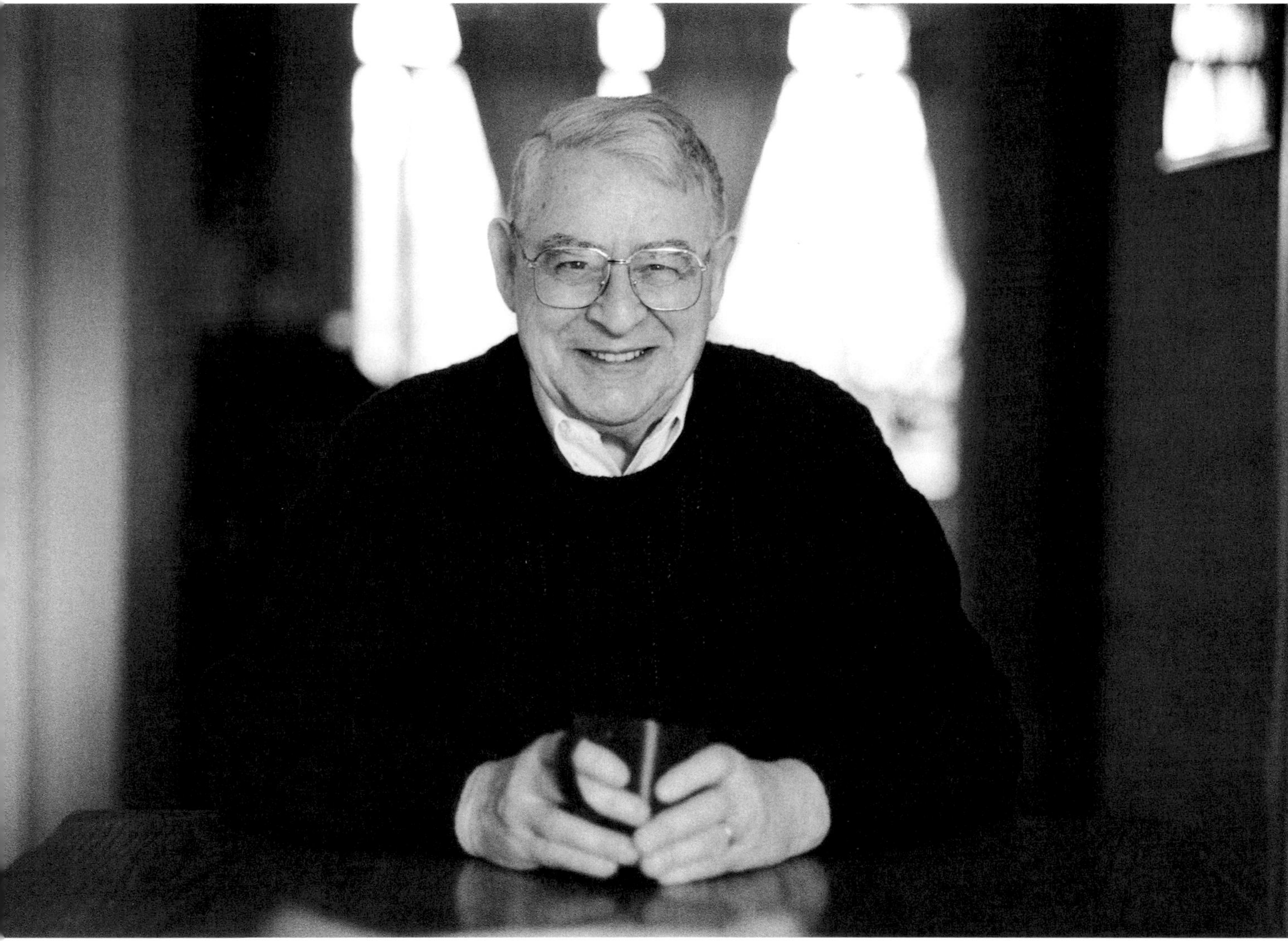

Ivar Giaever *1929

The Nobel Prize in Physics 1973 was divided, one half jointly to Leo Esaki and Ivar Giaever "for their experimental discoveries regarding tunneling phenomena in semiconductors and superconductors, respectively" and the other half to Brian David Josephson "for his theoretical predictions of the properties of a supercurrent through a tunnel barrier, in particular those phenomena which are generally known as the Josephson effects."

This tongue is more than
relatively famous –
but where does it come from?

Baden-Württemberg

Vitaly L. Ginzburg *1916 †2009

The Nobel Prize in Physics 2003 was awarded jointly to Alexei A. Abrikosov, Vitaly L. Ginzburg and Anthony J. Leggett "for pioneering contributions to the theory of superconductors and superfluids."

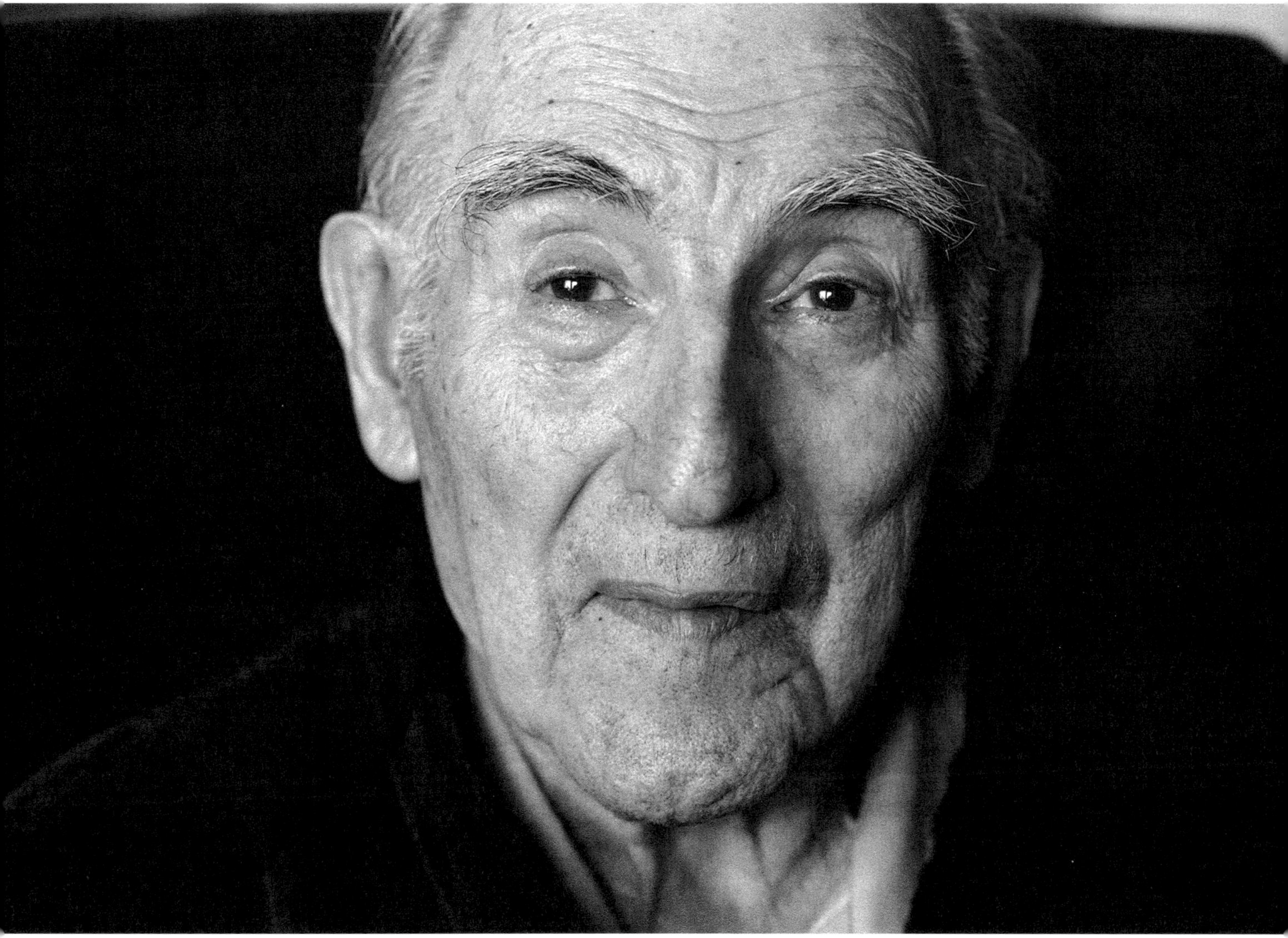

Donald Arthur Glaser *1926   †2013

The Nobel Prize in Physics 1960 was awarded to Donald A. Glaser "for the invention of the bubble chamber."

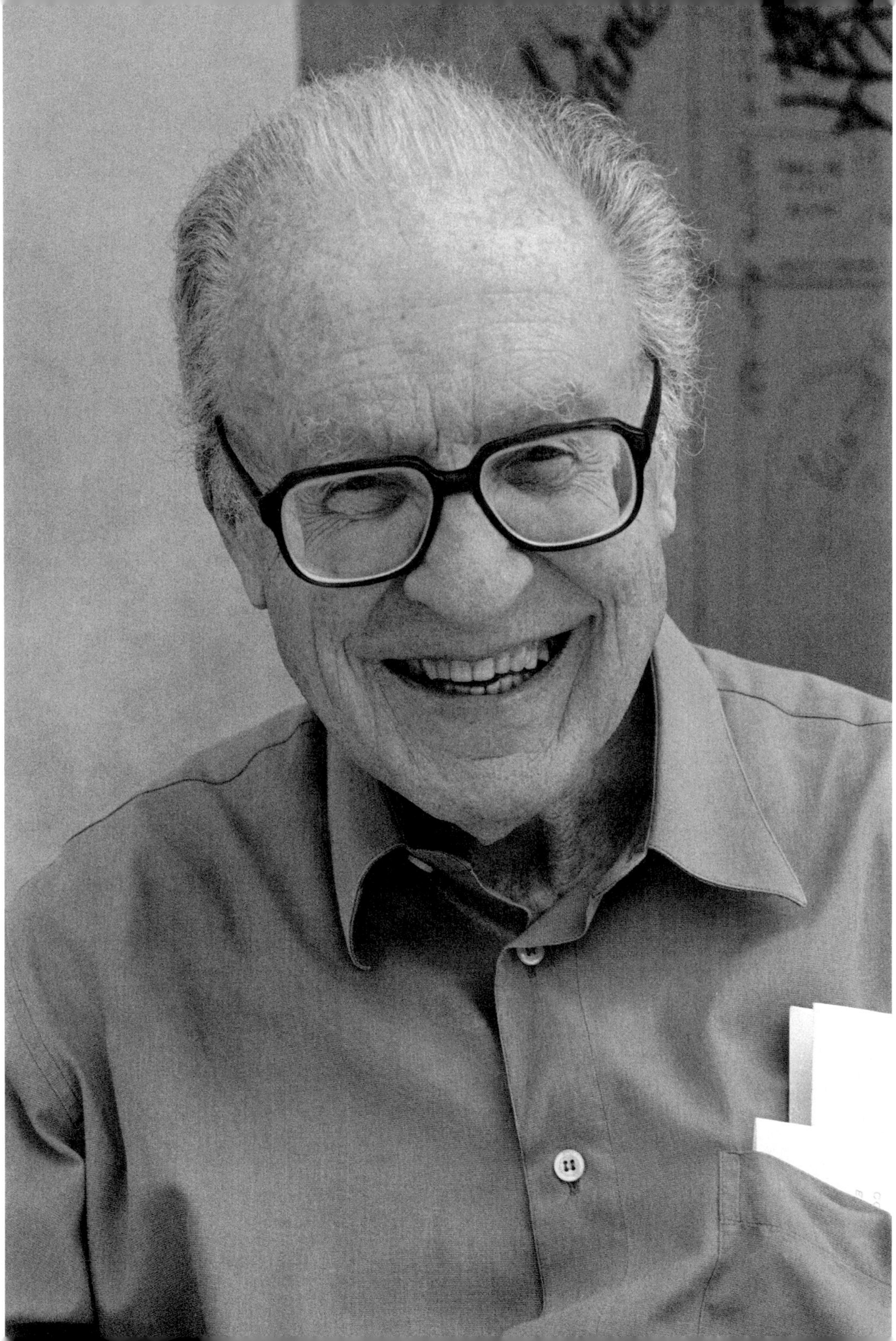

Sheldon Lee Glashow *1932

The Nobel Prize in Physics 1979 was awarded jointly to Sheldon Lee Glashow, Abdus Salam and Steven Weinberg "for their contributions to the theory of the unified weak and electromagnetic interaction between elementary particles, including, inter alia, the prediction of the weak neutral current."

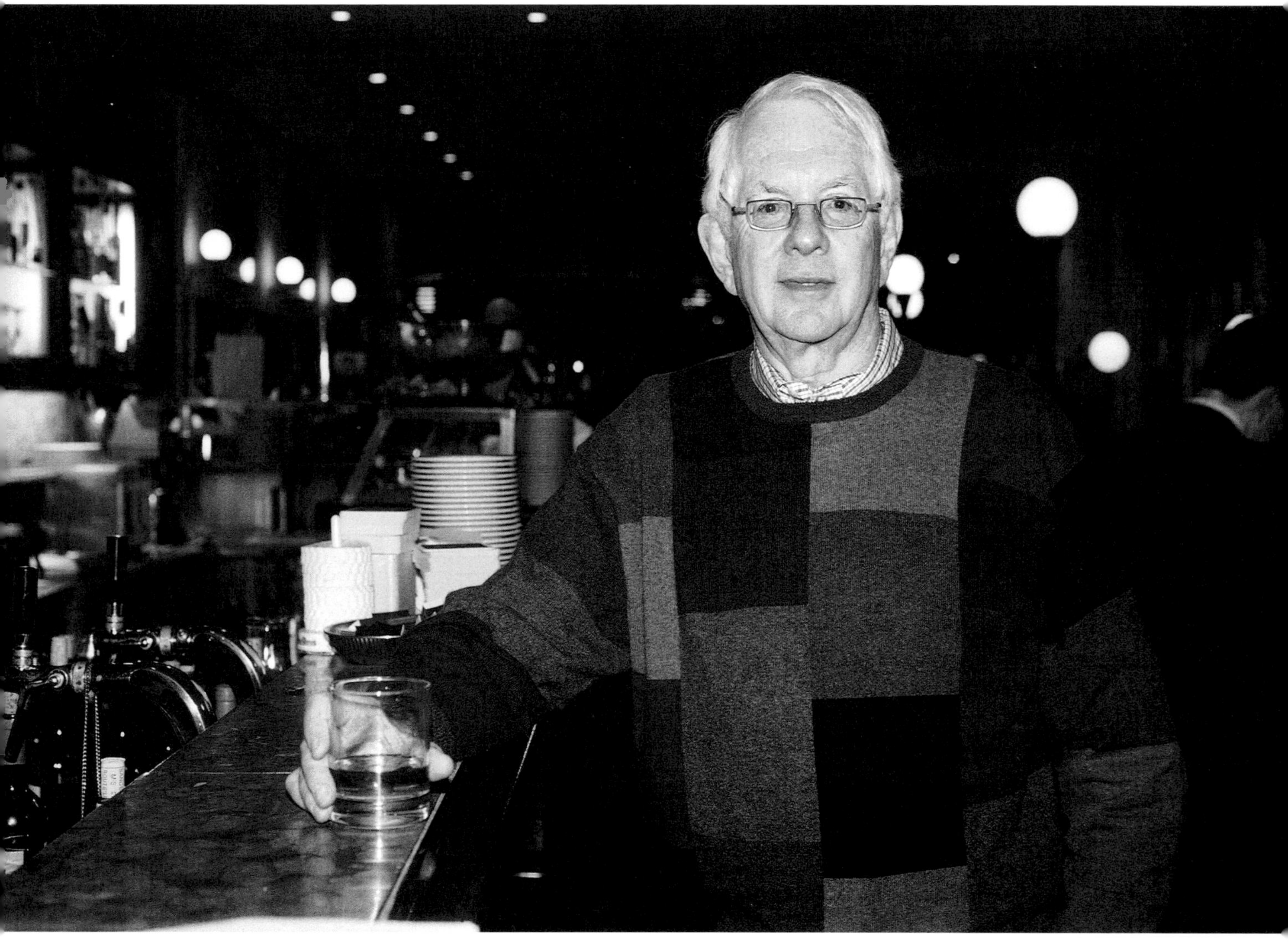

Roy J. Glauber *1925 †2018

The Nobel Prize in Physics 2005 was divided, one half awarded to Roy J. Glauber "for his contribution to the quantum theory of optical coherence", the other half jointly to John L. Hall and Theodor W. Hänsch "for their contributions to the development of laser-based precision spectroscopy, including the optical frequency comb technique."

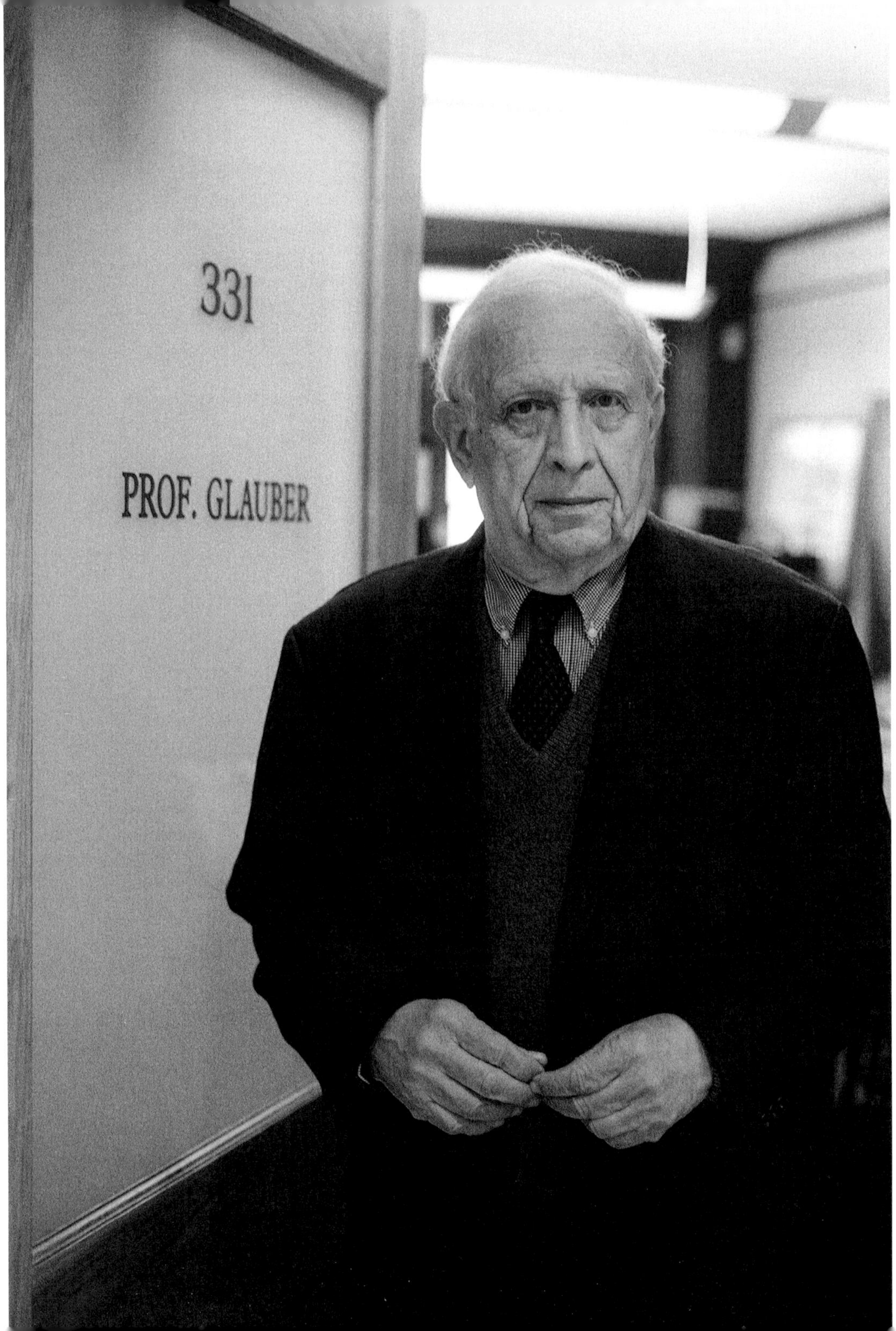
331
PROF. GLAUBER

David J. Gross *1941

The Nobel Prize in Physics 2004 was awarded jointly to David J. Gross, H. David Politzer and Frank Wilczek "for the discovery of asymptotic freedom in the theory of the strong interaction."

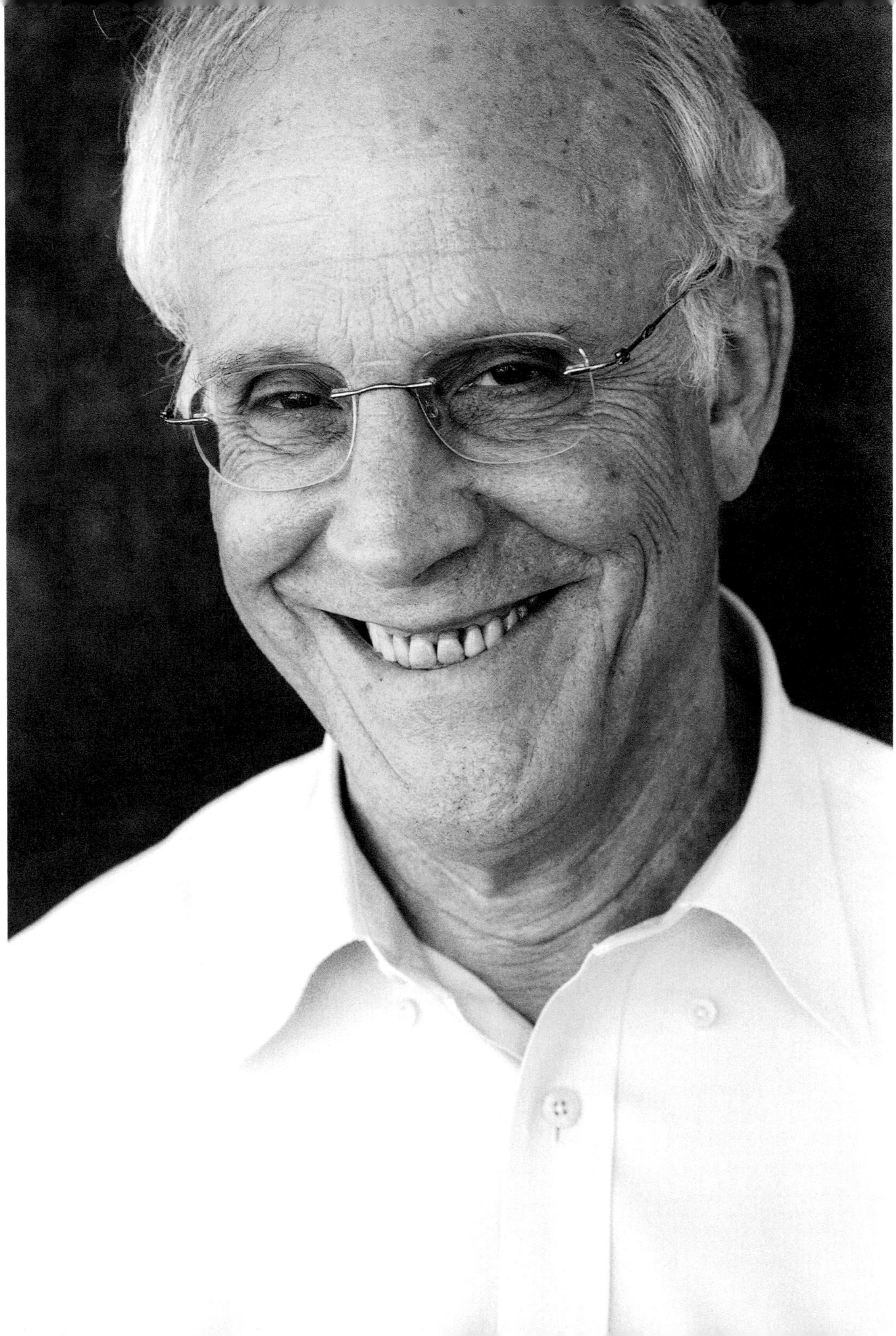

Peter Grünberg *1939 †2018

The Nobel Prize in Physics 2007 was awarded jointly to Albert Fert and Peter Grünberg "for the discovery of Giant Magnetoresistance."

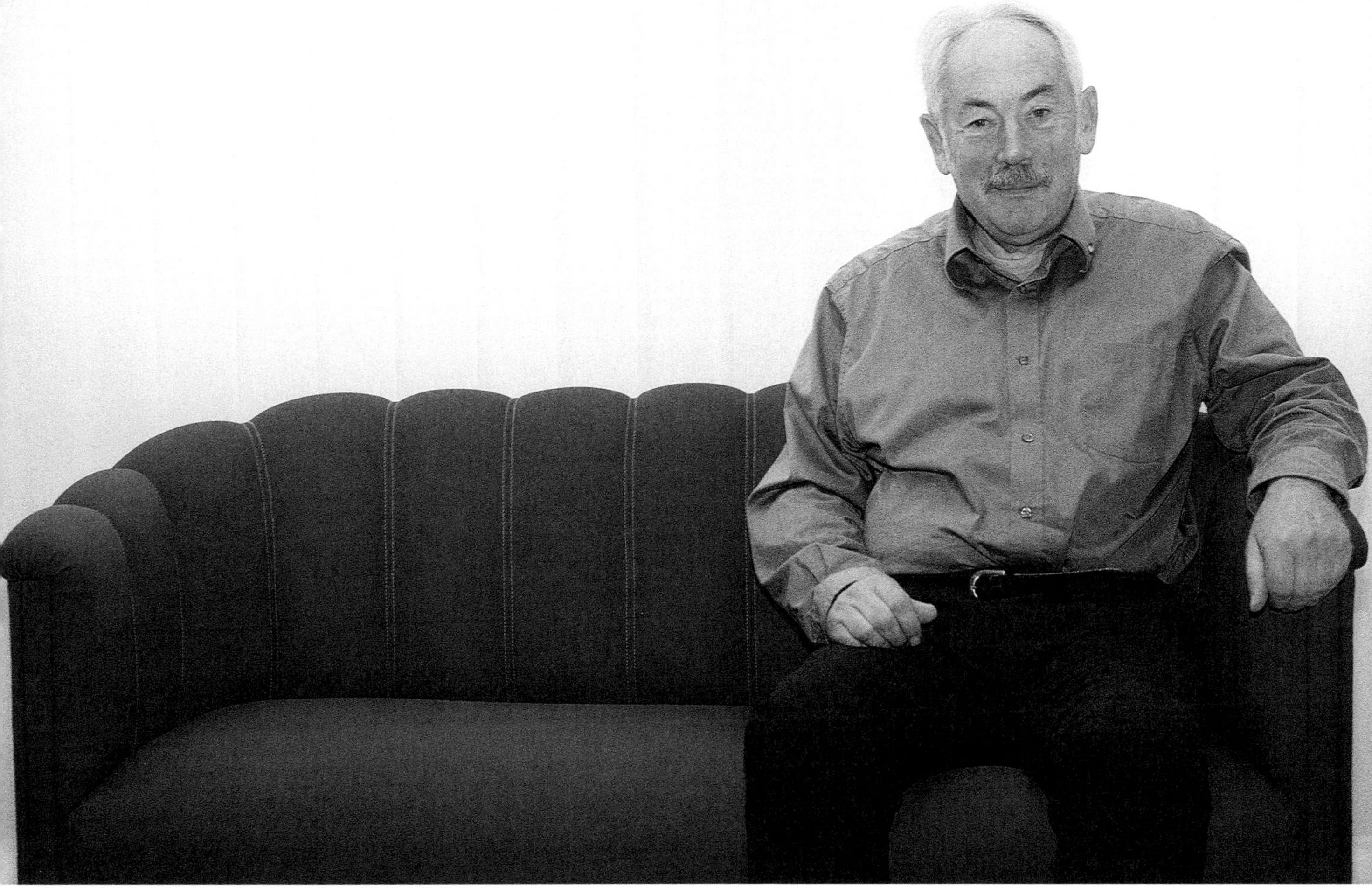

F. Duncan M. Haldane *1951

The Nobel Prize in Physics 2016 was divided, one half awarded to David J. Thouless, the other half jointly to F. Duncan M. Haldane and J. Michael Kosterlitz "for theoretical discoveries of topological phase transitions and topological phases of matter."

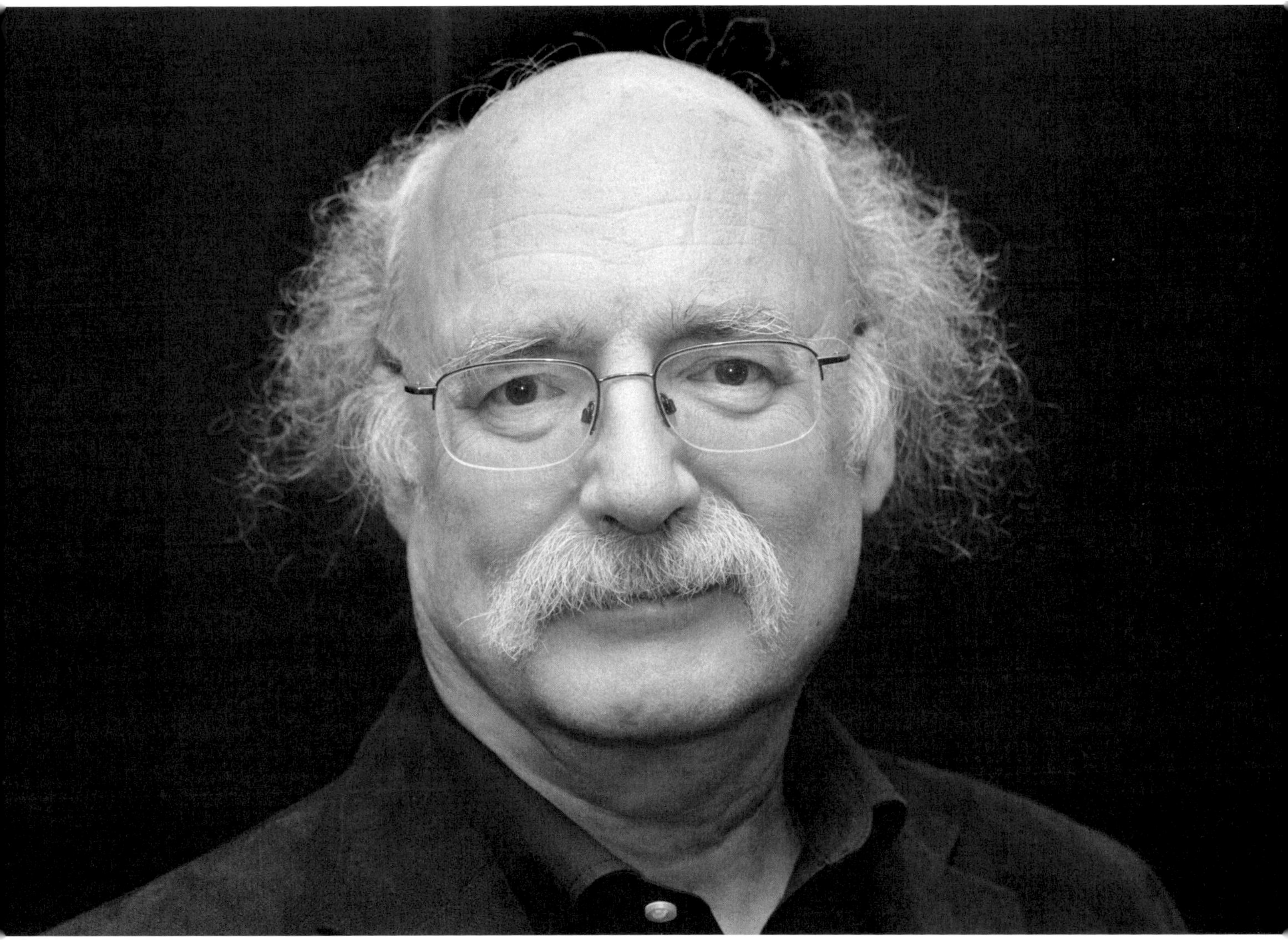

John L. Hall *1934

The Nobel Prize in Physics 2005 was divided, one half awarded to Roy J. Glauber "for his contribution to the quantum theory of optical coherence", the other half jointly to John L. Hall and Theodor W. Hänsch "for their contributions to the development of laser-based precision spectroscopy, including the optical frequency comb technique."

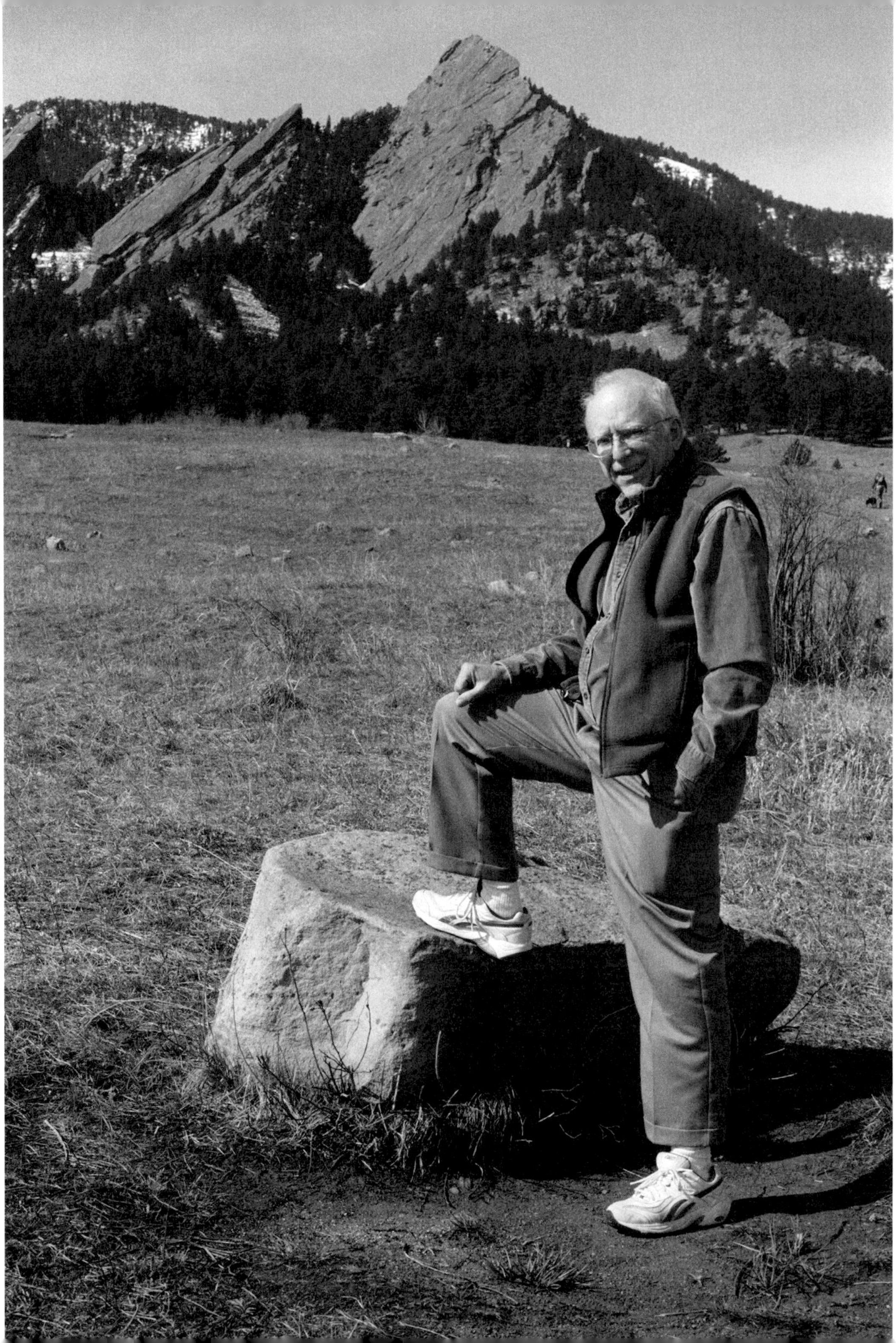

Theodor W. Hänsch *1941

The Nobel Prize in Physics 2005 was divided, one half awarded to Roy J. Glauber "for his contribution to the quantum theory of optical coherence", the other half jointly to John L. Hall and Theodor W. Hänsch "for their contributions to the development of laser-based precision spectroscopy, including the optical frequency comb technique."

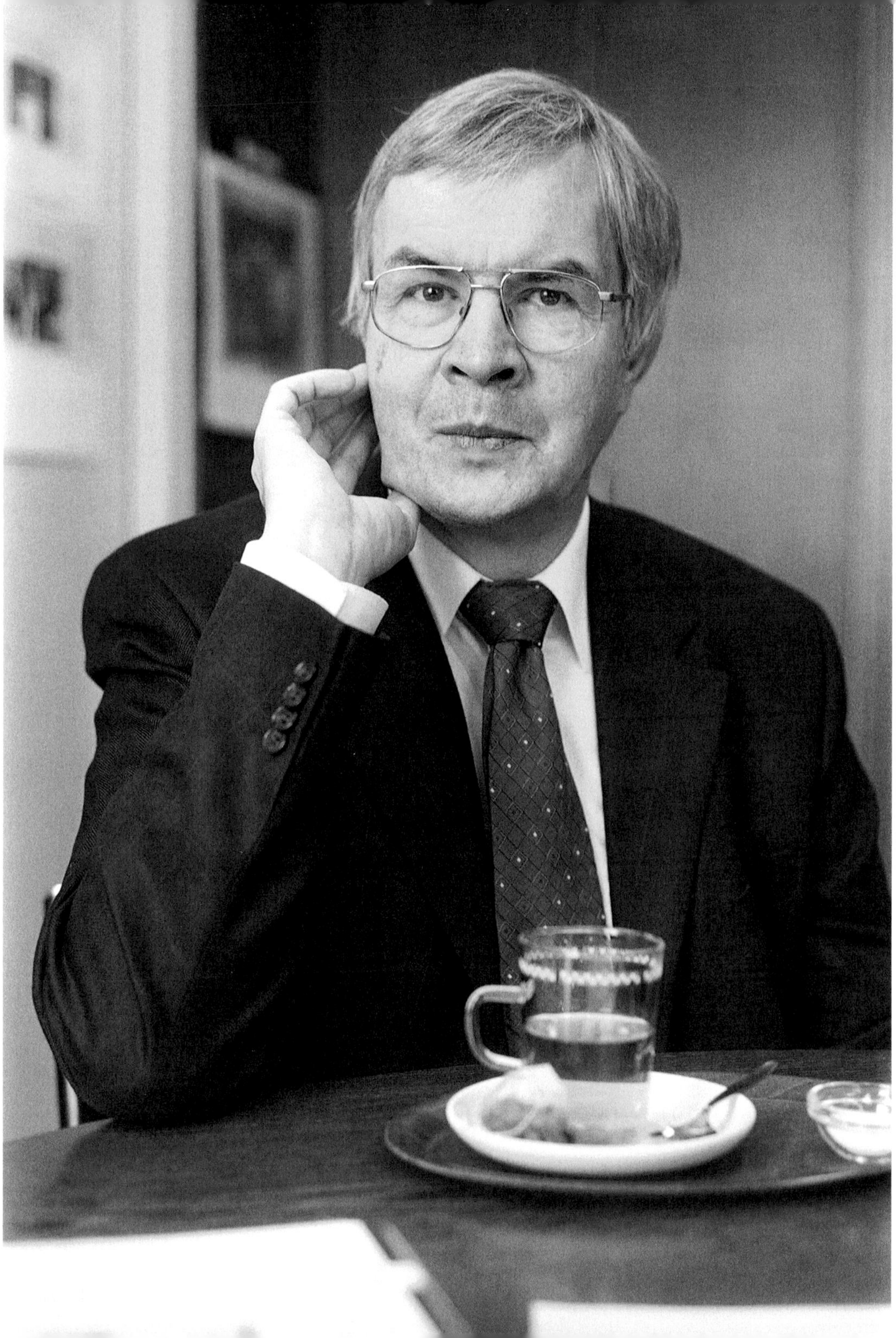

Serge Haroche *1944

The Nobel Prize in Physics 2012 was awarded jointly to Serge Haroche and David J. Wineland "for ground-breaking experimental methods that enable measuring and manipulation of individual quantum systems."

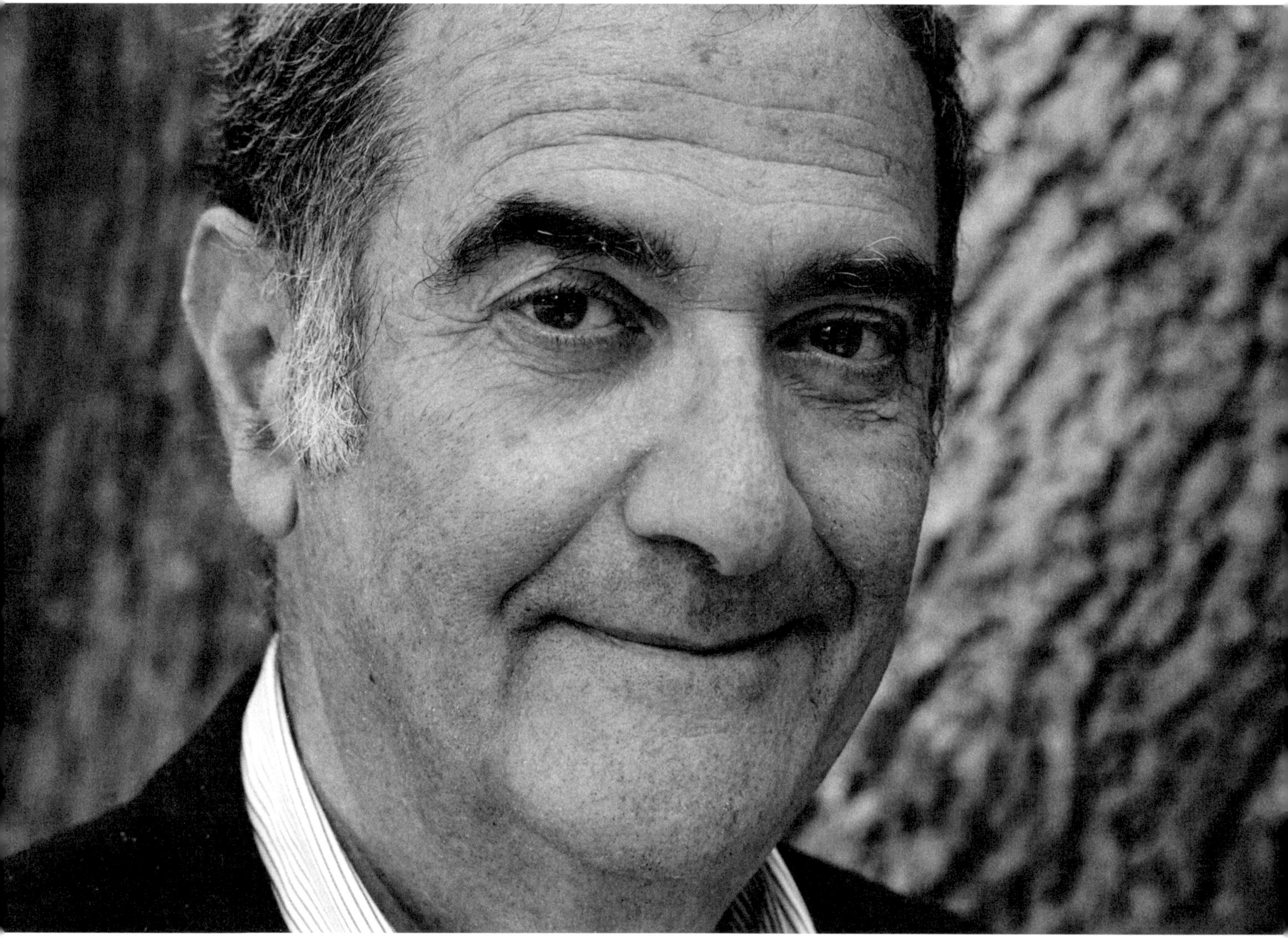

Antony Hewish *1924

The Nobel Prize in Physics 1974 was awarded jointly to Sir Martin Ryle and Antony Hewish "for their pioneering research in radio astrophysics: Ryle for his observations and inventions, in particular of the aperture synthesis technique, and Hewish for his decisive role in the discovery of pulsars."

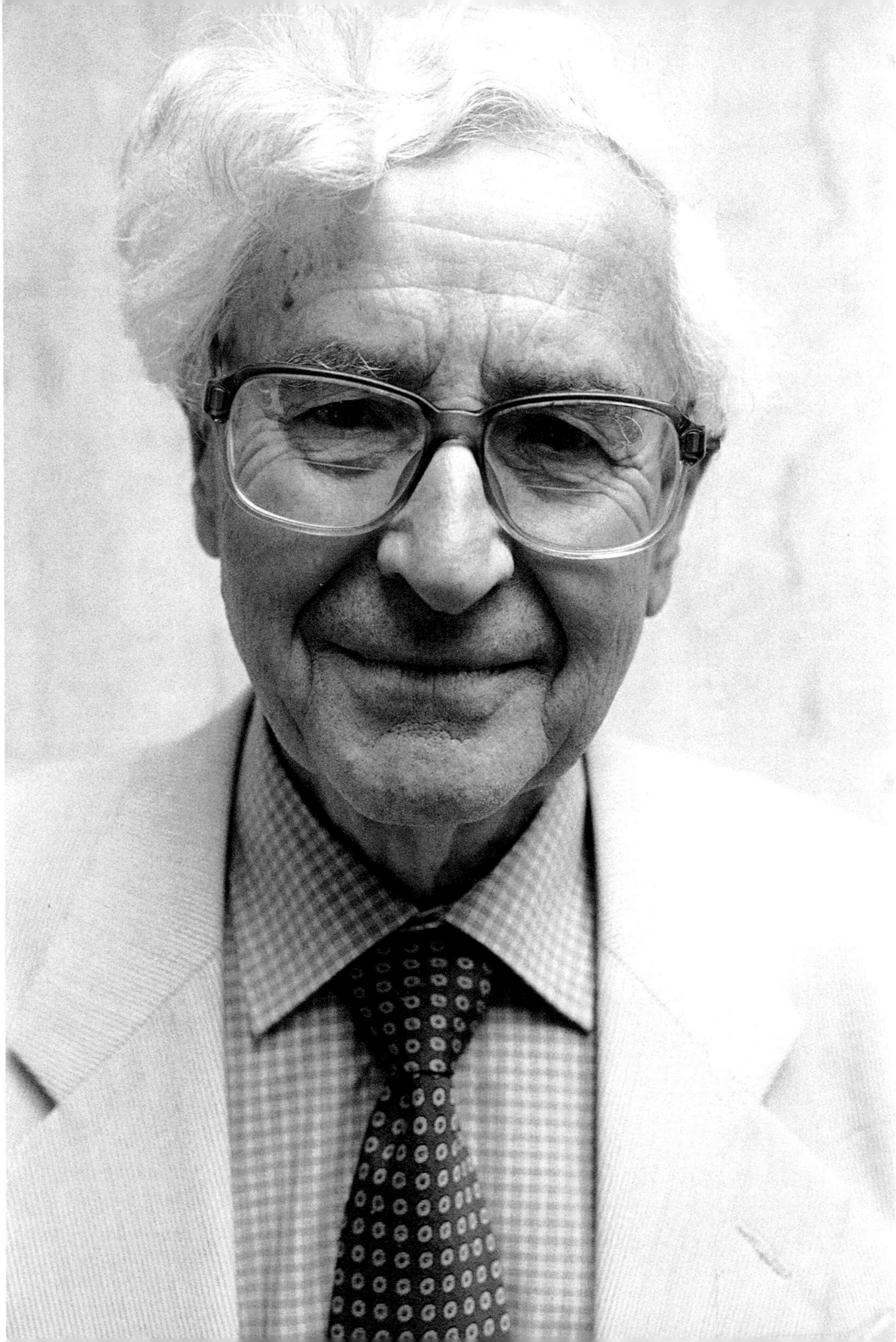

Peter W. Higgs *1929

The Nobel Prize in Physics 2013 was awarded jointly to François Englert and Peter W. Higgs "for the theoretical discovery of a mechanism that contributes to our understanding of the origin of mass of subatomic particles, and which recently was confirmed through the discovery of the predicted fundamental particle, by the ATLAS and CMS experiments at CERN's Large Hadron Collider."

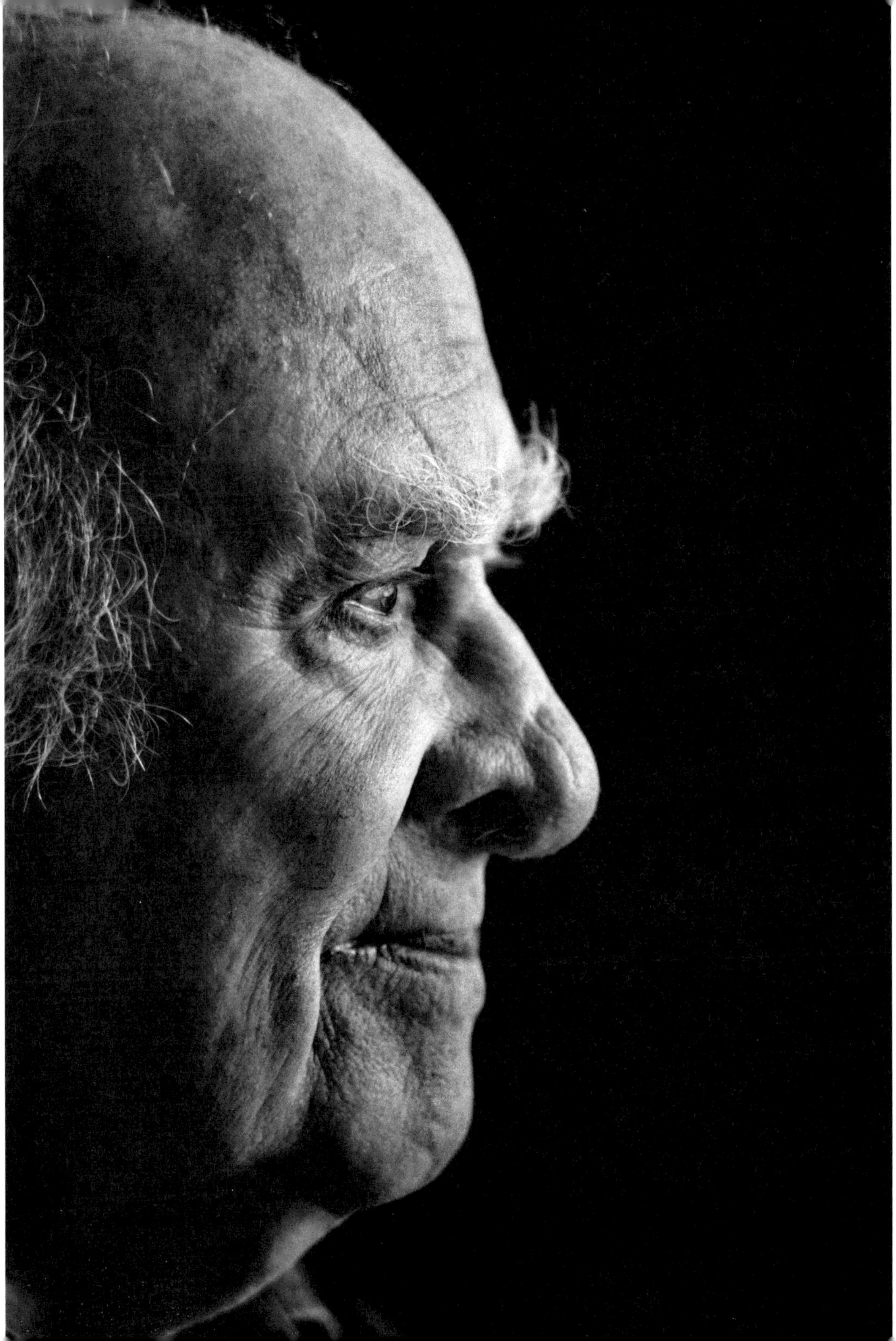

Gerardus 't Hooft *1946

The Nobel Prize in Physics 1999 was awarded jointly to Gerardus 't Hooft and Martinus J. G. Veltman "for elucidating the quantum structure of electroweak interactions in physics."

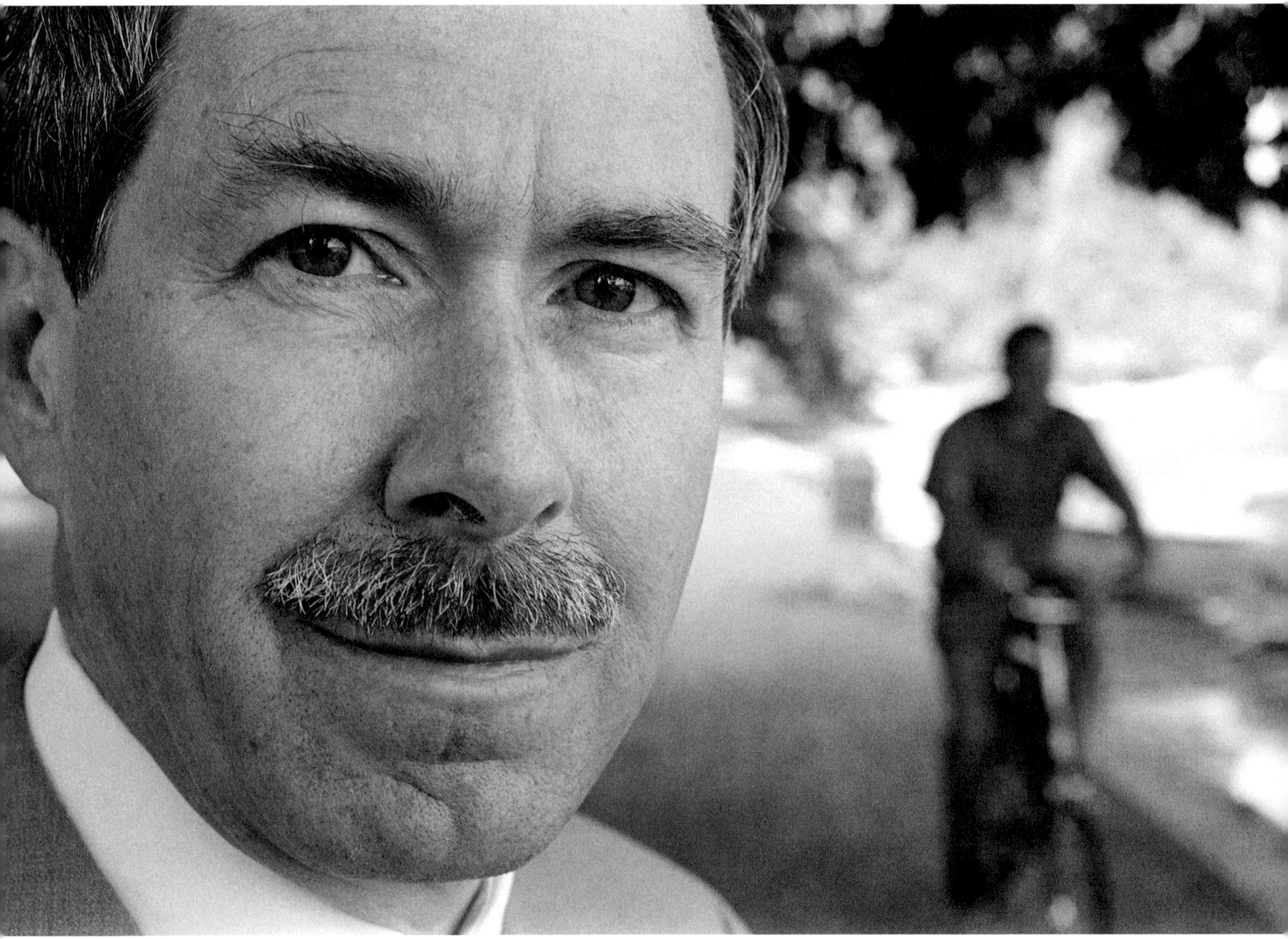

Russell A. Hulse *1950

The Nobel Prize in Physics 1993 was awarded jointly to Russell A. Hulse and Joseph H. Taylor Jr. "for the discovery of a new type of pulsar, a discovery that has opened up new possibilities for the study of gravitation."

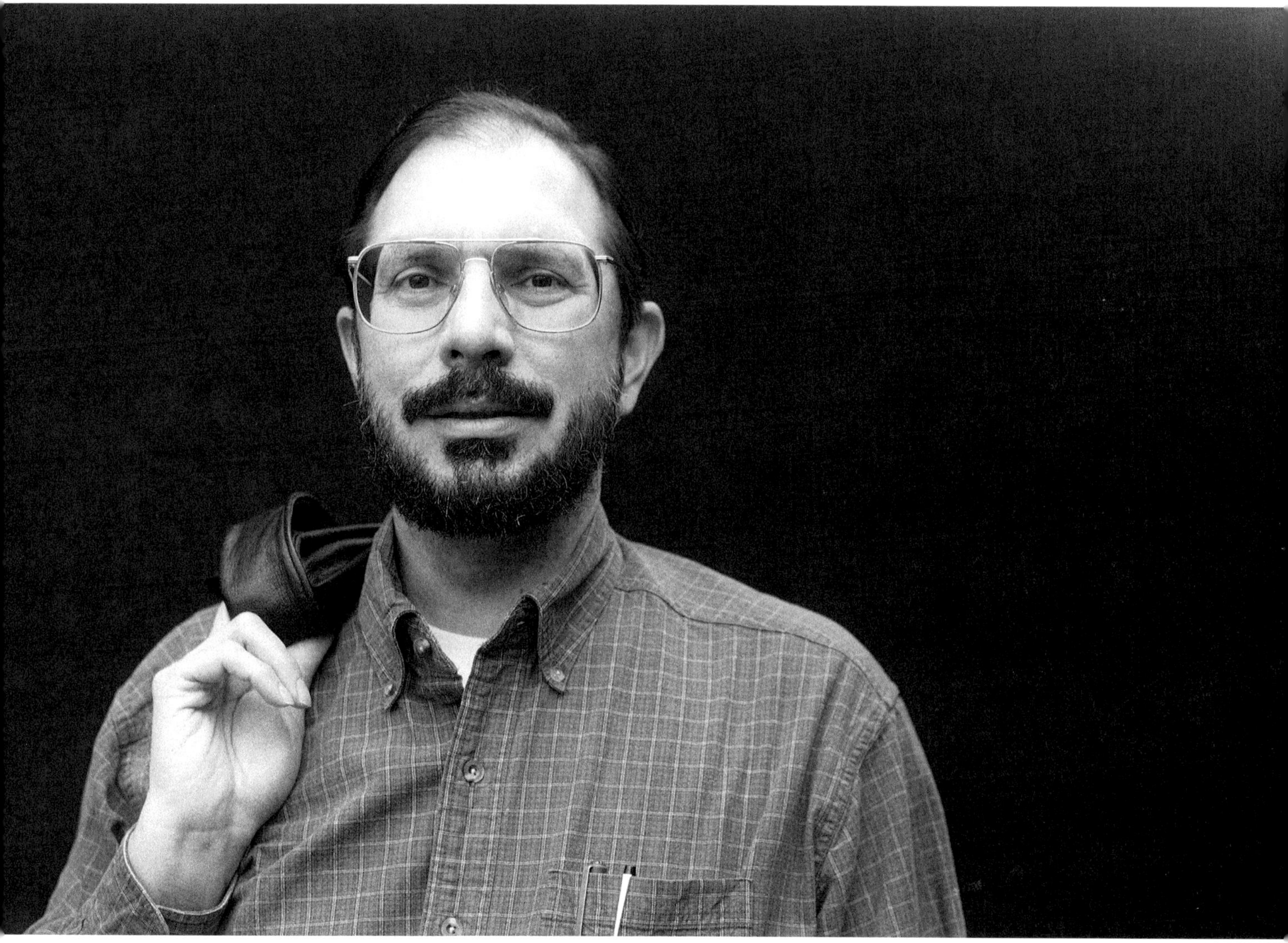

Brian David Josephson *1940

The Nobel Prize in Physics 1973 was divided, one half jointly to Leo Esaki and Ivar Giaever "for their experimental discoveries regarding tunneling phenomena in semiconductors and superconductors, respectively" and the other half to Brian David Josephson "for his theoretical predictions of the properties of a supercurrent through a tunnel barrier, in particular those phenomena which are generally known as the Josephson effects."

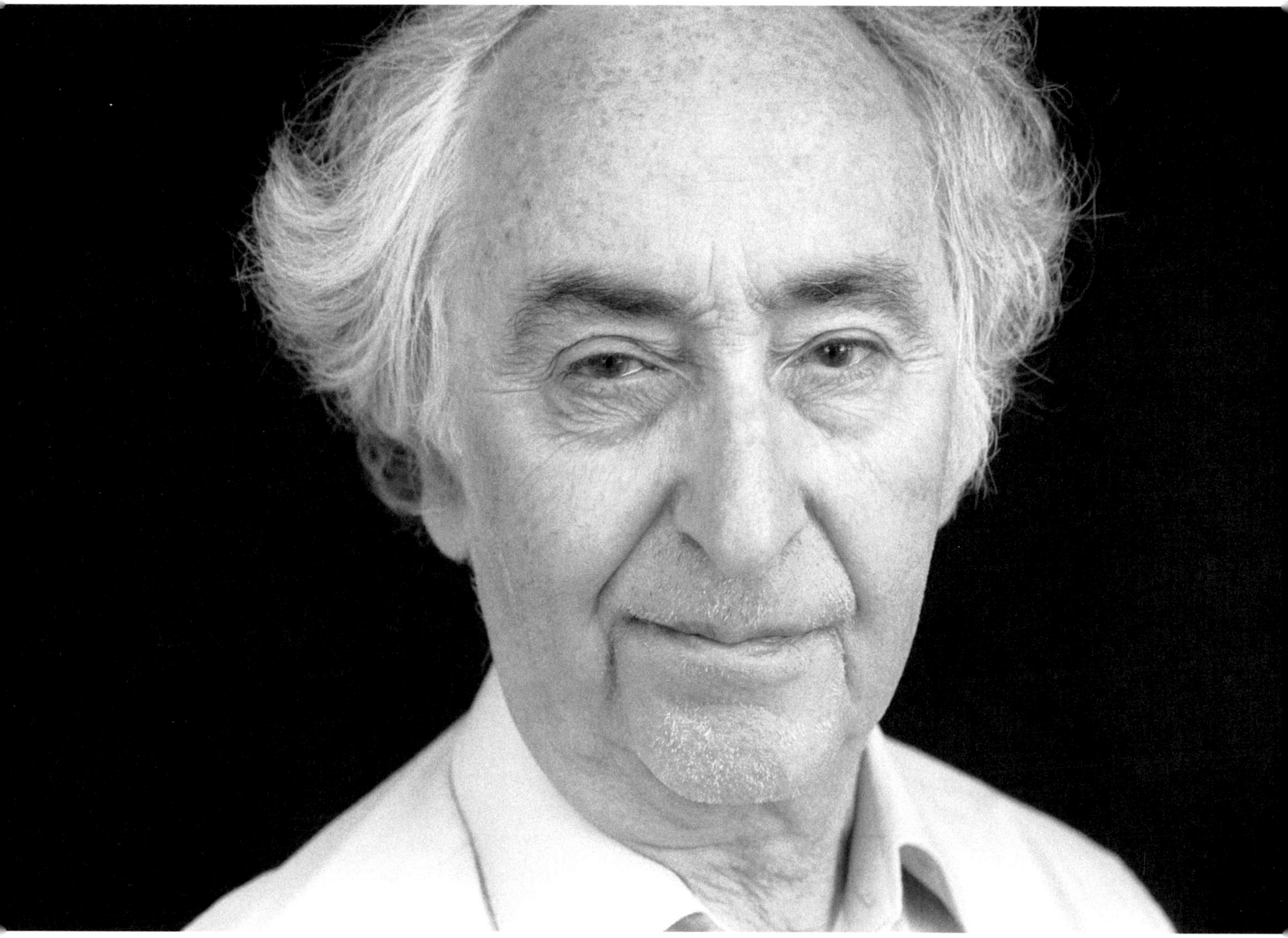

Takaaki Kajita *1959

The Nobel Prize in Physics 2015 was awarded jointly to Takaaki Kajita and Arthur B. McDonald "for the discovery of neutrino oscillations, which shows that neutrinos have mass."

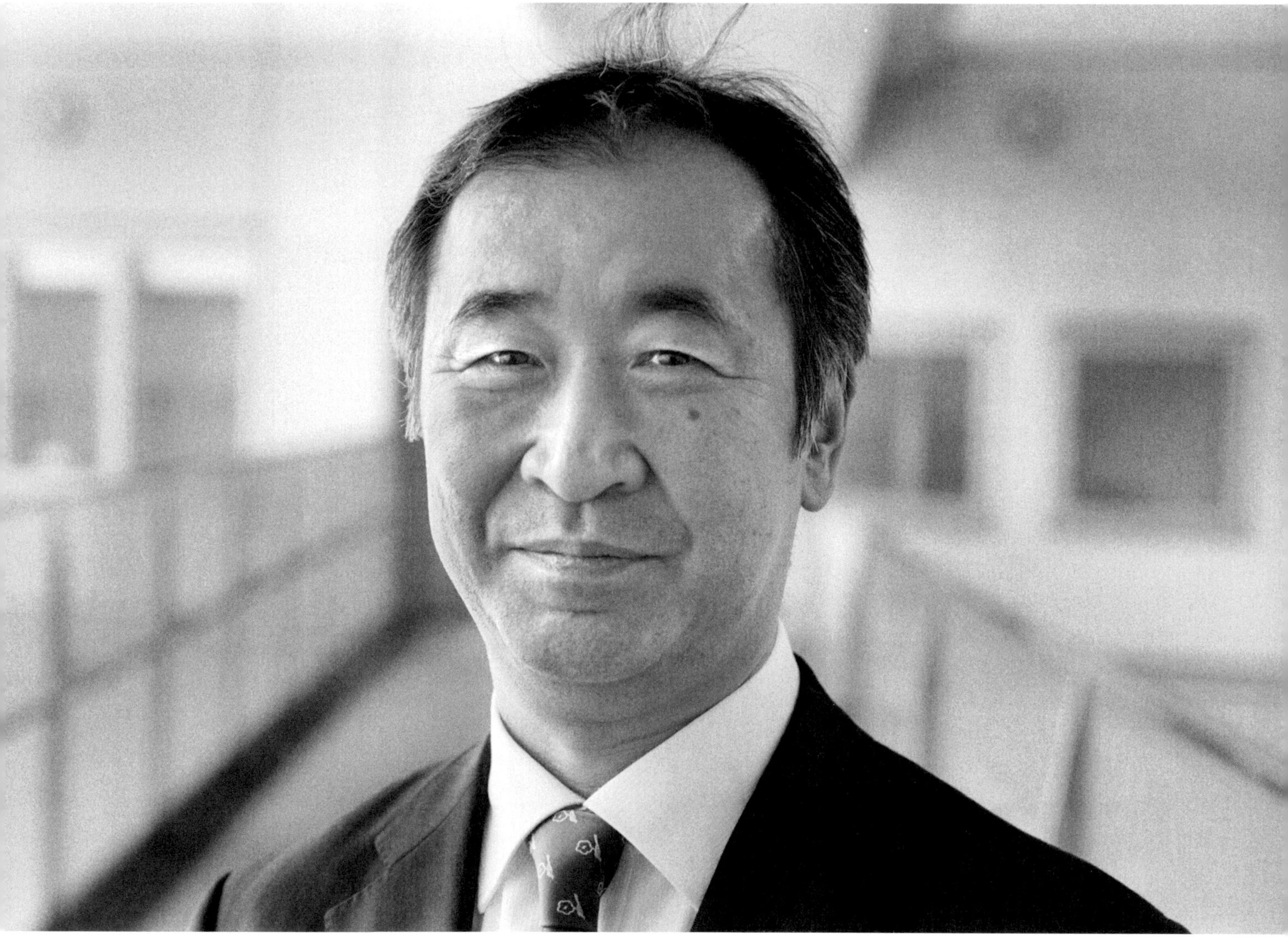

Charles Kuen Kao *1933 †2018

The Nobel Prize in Physics 2009 was divided, one half awarded to Charles Kuen Kao "for groundbreaking achievements concerning the transmission of light in fibers for optical communication", the other half jointly to Willard S. Boyle and George E. Smith "for the invention of an imaging semiconductor circuit – the CCD sensor."

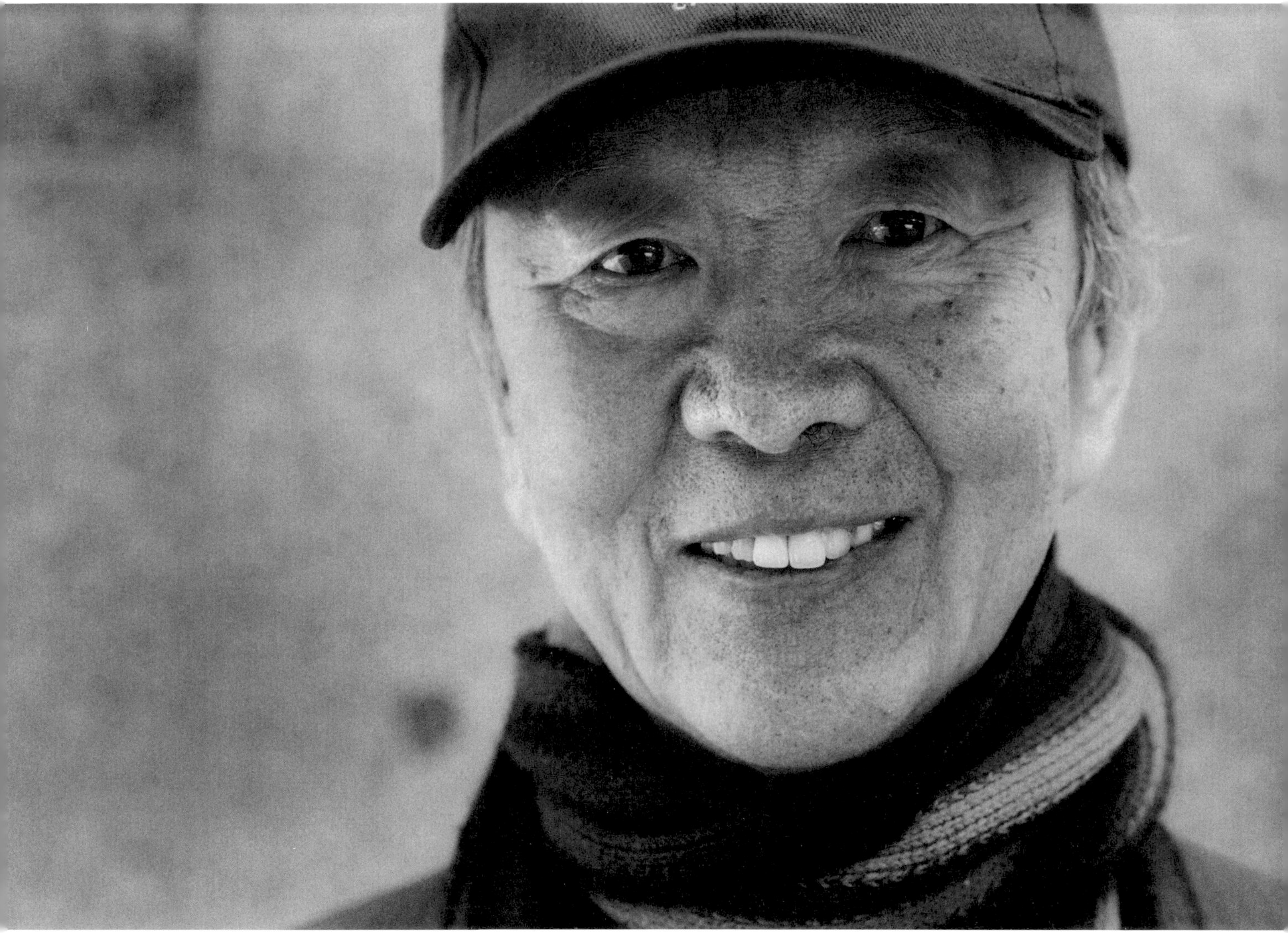

Wolfgang Ketterle *1957

The Nobel Prize in Physics 2001 was awarded jointly to Eric A. Cornell, Wolfgang Ketterle and Carl E. Wieman "for the achievement of Bose-Einstein condensation in dilute gases of alkali atoms, and for early fundamental studies of the properties of the condensates."

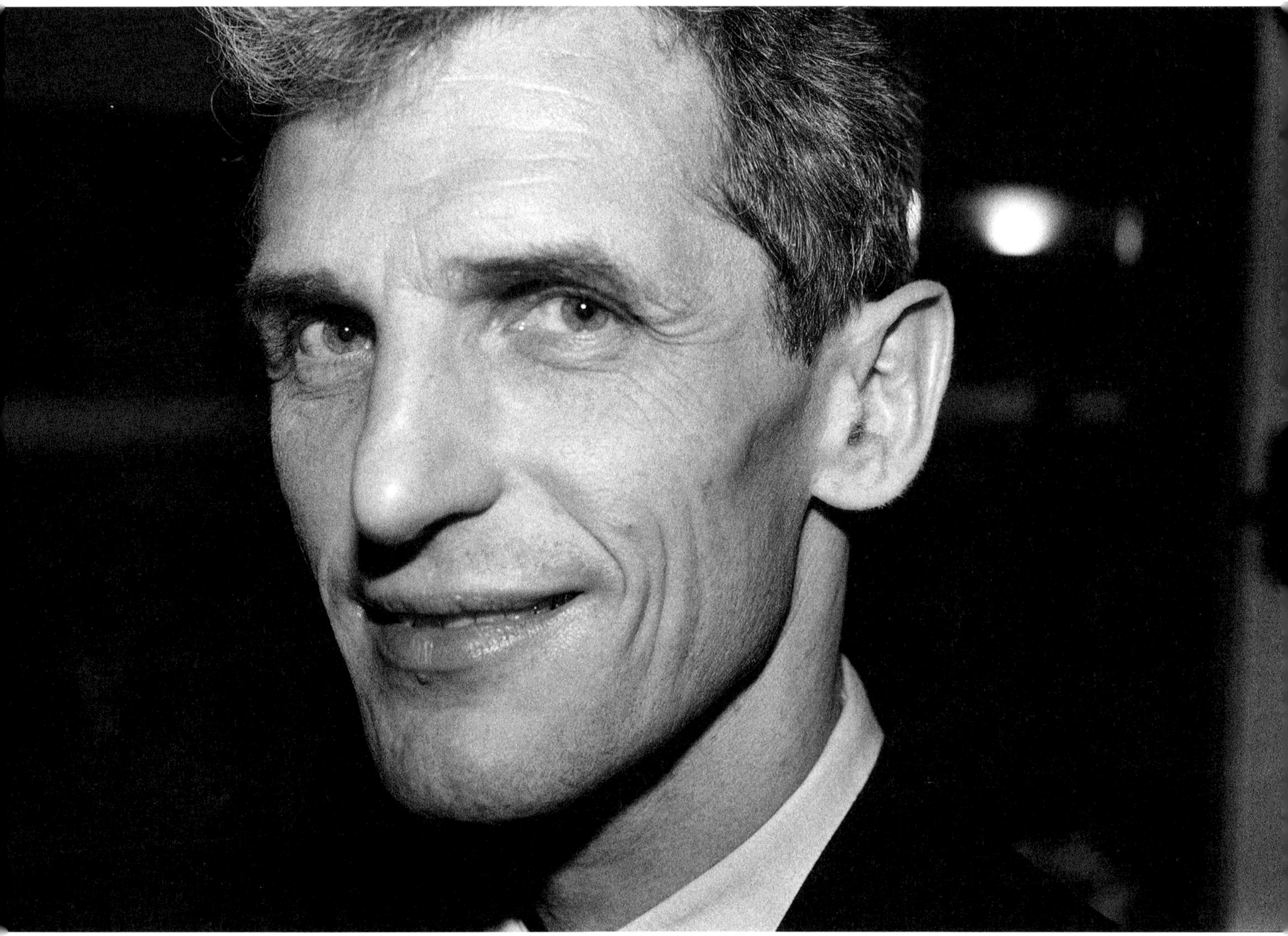

Jack S. Kilby *1923   †2005

The Nobel Prize in Physics 2000 was awarded "for basic work on information and communication technology" with one half jointly to Zhores I. Alferov and Herbert Kroemer "for developing semi-conductor heterostructures used in high-speed- and opto-electronics" and the other half to Jack S. Kilby "for his part in the invention of the integrated circuit."

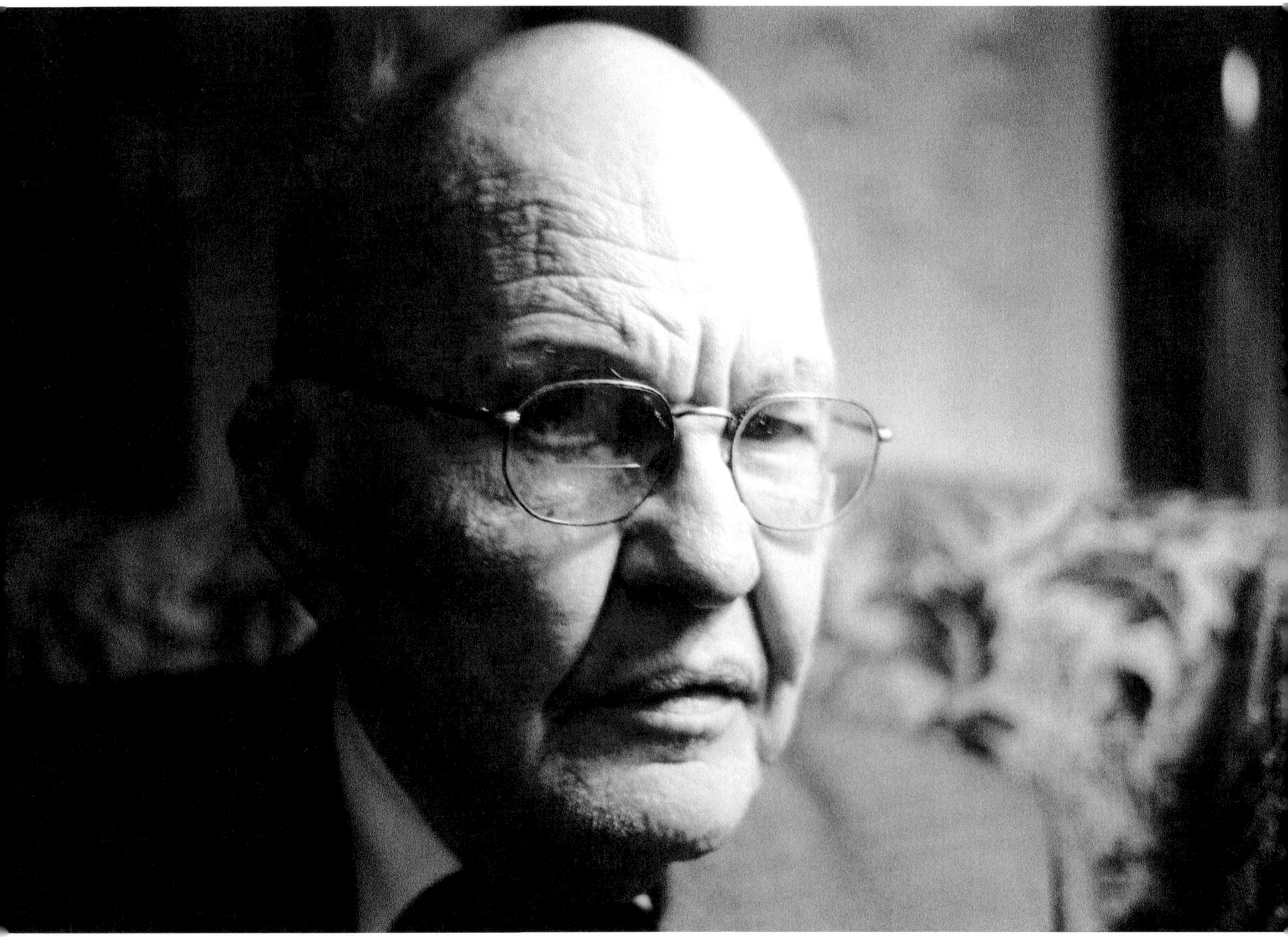

Klaus von Klitzing *1943

The Nobel Prize in Physics 1985 was awarded to Klaus von Klitzing "for the discovery of the quantized Hall effect."

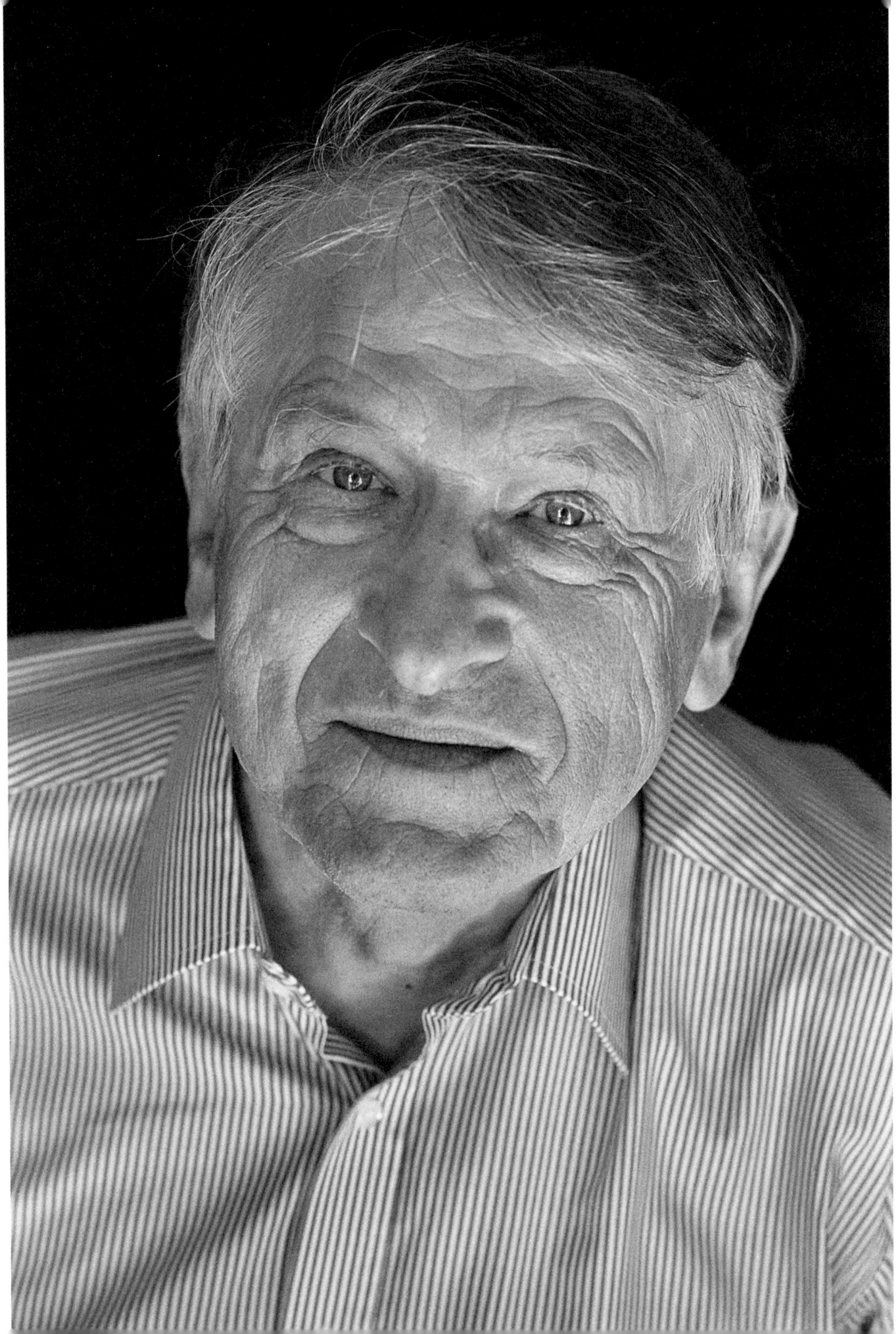

Makoto Kobayashi *1944

The Nobel Prize in Physics 2008 was divided, one half awarded to Yoichiro Nambu "for the discovery of the mechanism of spontaneous broken symmetry in subatomic physics", the other half jointly to Makoto Kobayashi and Toshihide Maskawa "for the discovery of the origin of the broken symmetry which predicts the existence of at least three families of quarks in nature."

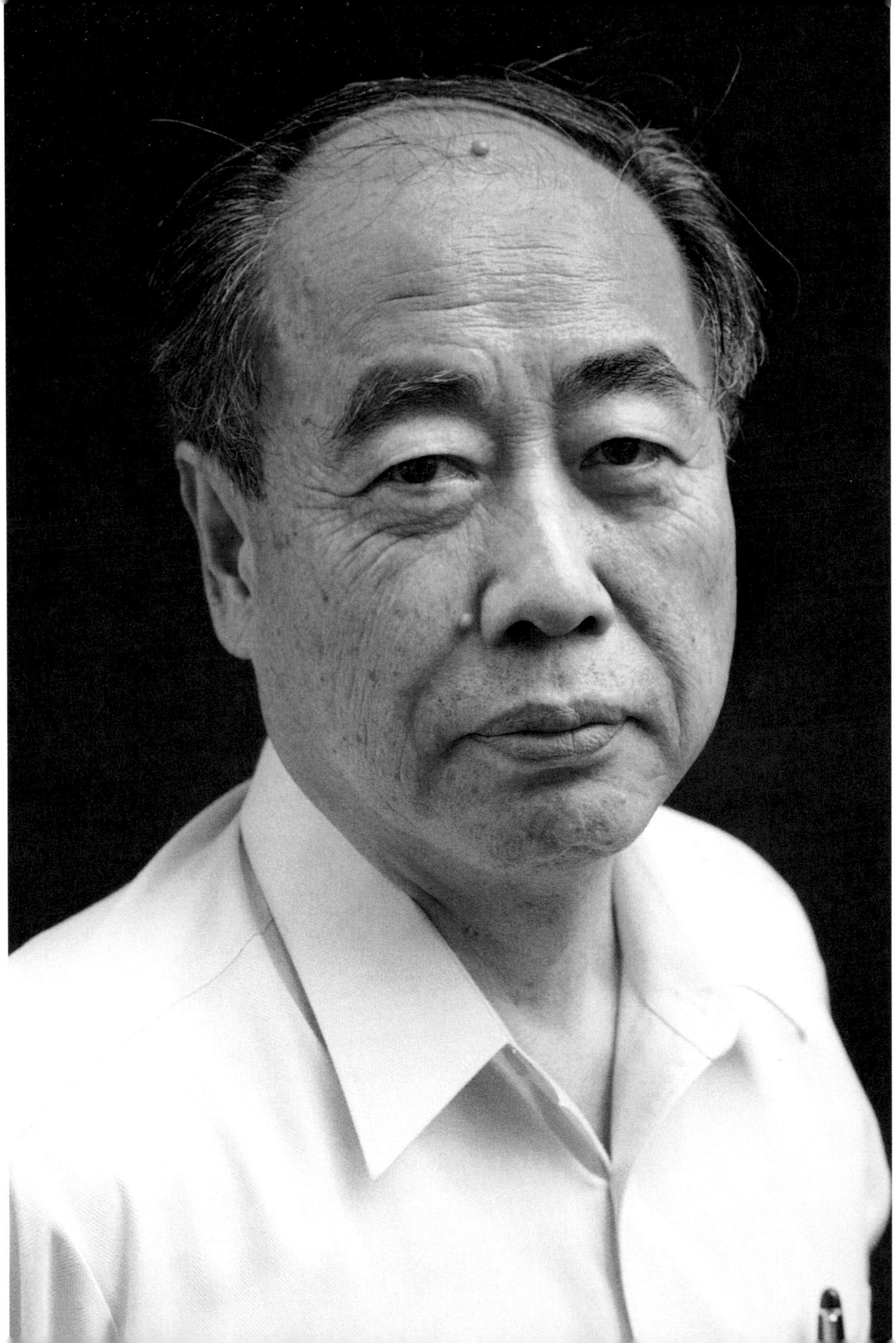

Masatoshi Koshiba *1926

The Nobel Prize in Physics 2002 was divided, one half jointly to Raymond Davis Jr. and Masatoshi Koshiba "for pioneering contributions to astrophysics, in particular for the detection of cosmic neutrinos" and the other half to Riccardo Giacconi "for pioneering contributions to astrophysics, which have led to the discovery of cosmic X-ray sources."

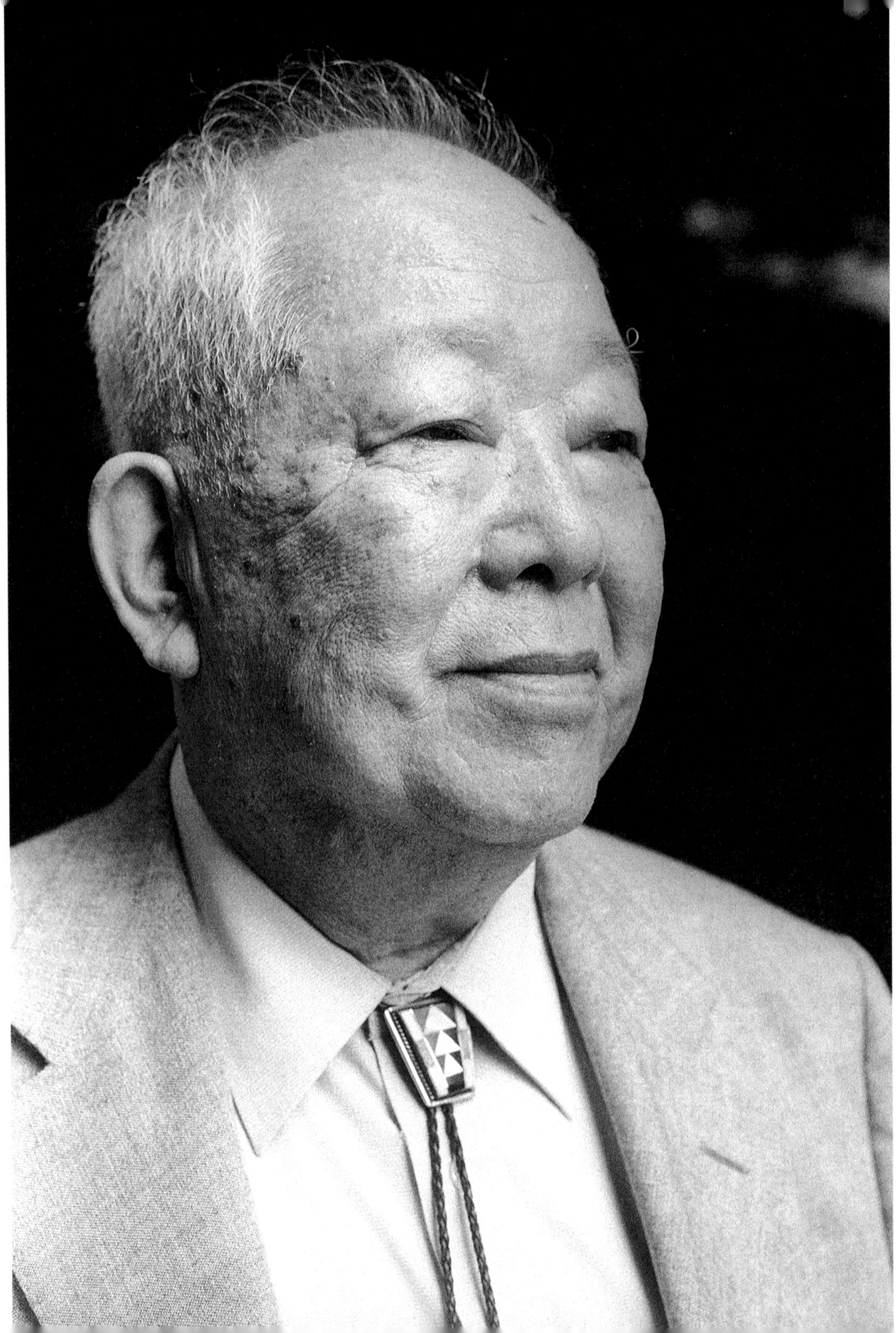

J. Michael Kosterlitz *1943

The Nobel Prize in Physics 2016 was divided, one half awarded to David J. Thouless, the other half jointly to F. Duncan M. Haldane and J. Michael Kosterlitz "for theoretical discoveries of topological phase transitions and topological phases of matter."

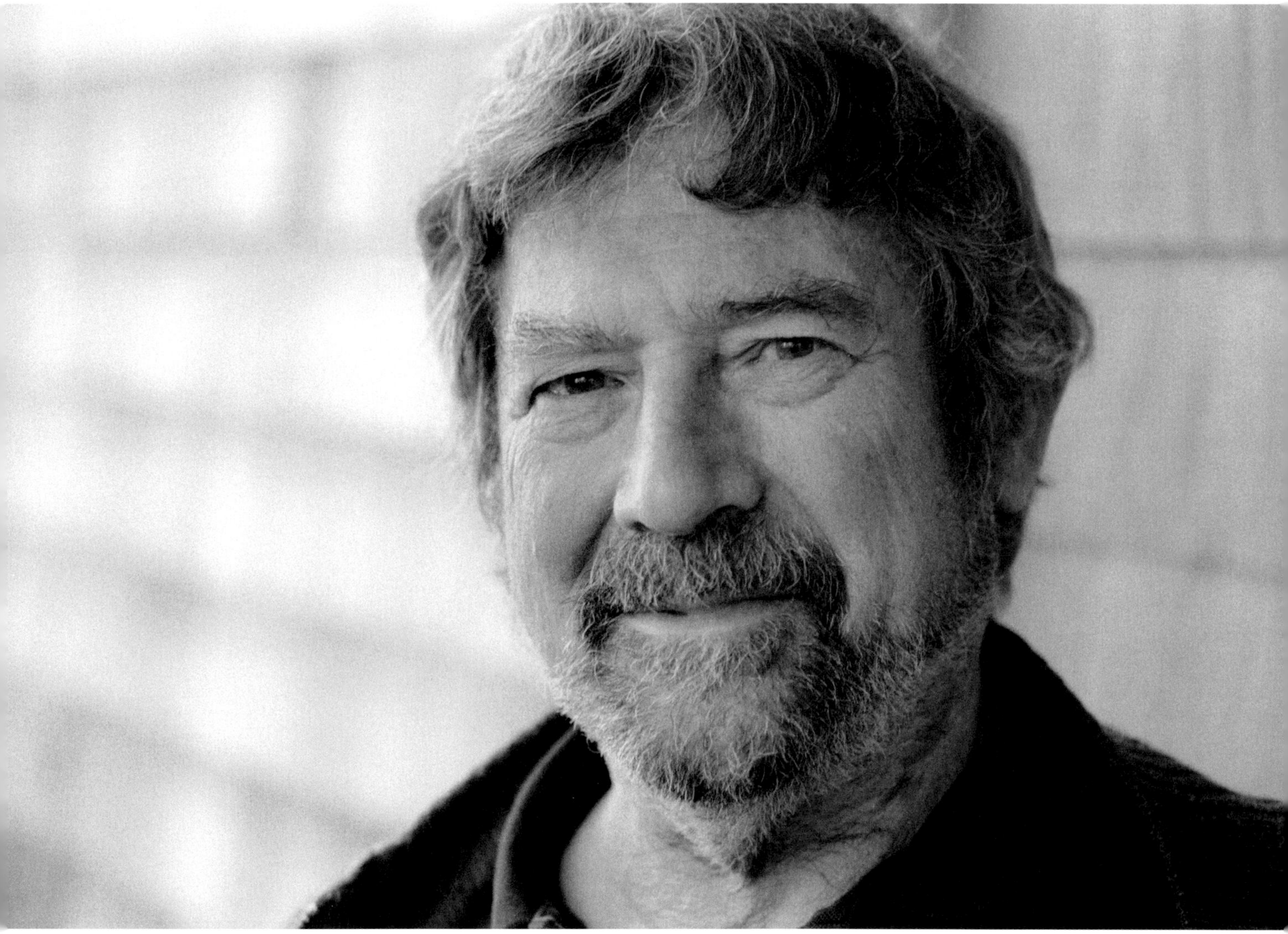

Herbert Kroemer *1928

The Nobel Prize in Physics 2000 was awarded "for basic work on information and communication technology" with one half jointly to Zhores I. Alferov and Herbert Kroemer "for developing semiconductor heterostructures used in high-speed- and opto-electronics" and the other half to Jack S. Kilby "for his part in the invention of the integrated circuit."

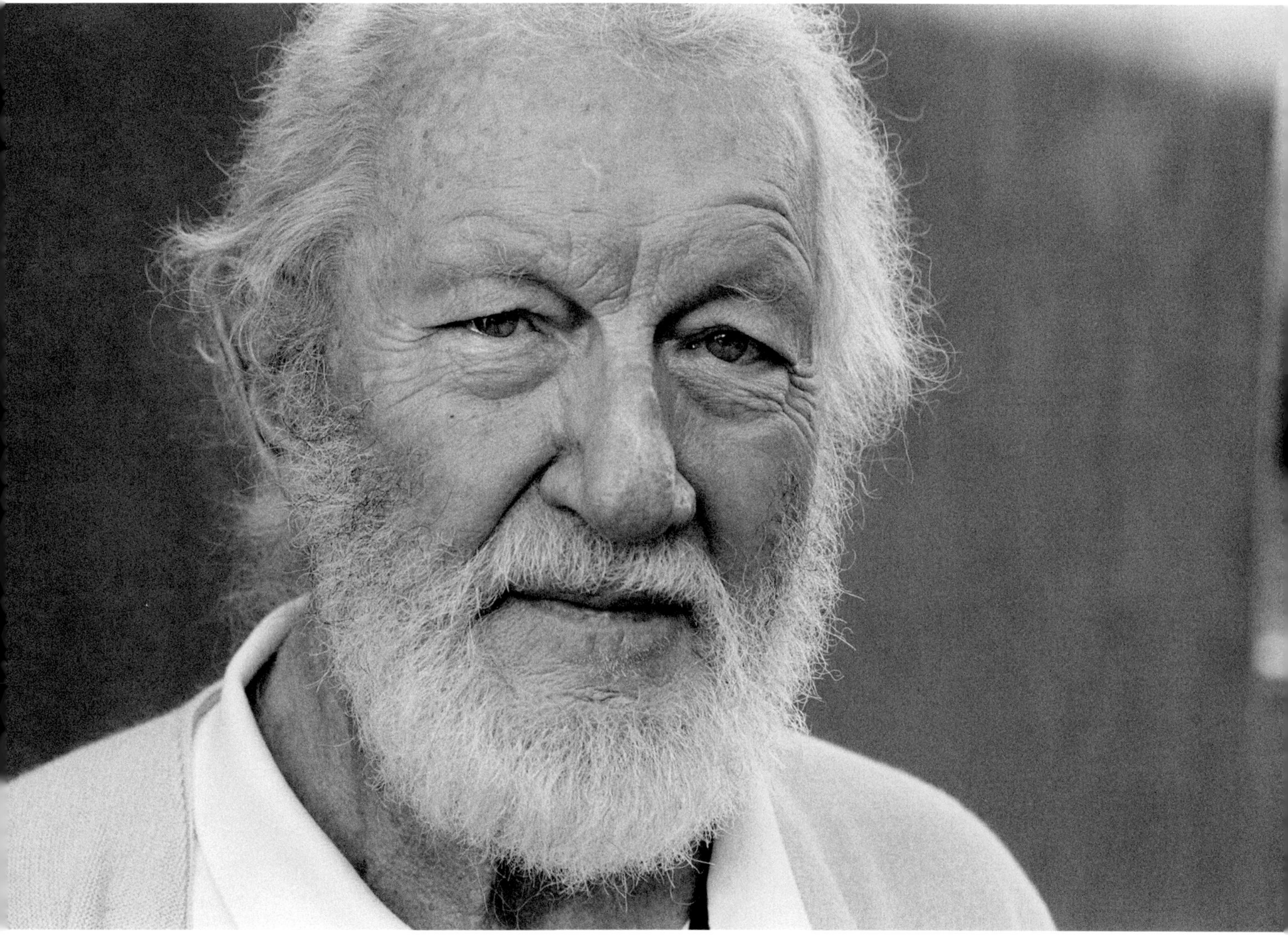

Willis Eugene Lamb *1913 †2008

The Nobel Prize in Physics 1955 was divided equally between Willis Eugene Lamb "for his discoveries concerning the fine structure of the hydrogen spectrum" and Polykarp Kusch "for his precision determination of the magnetic moment of the electron."

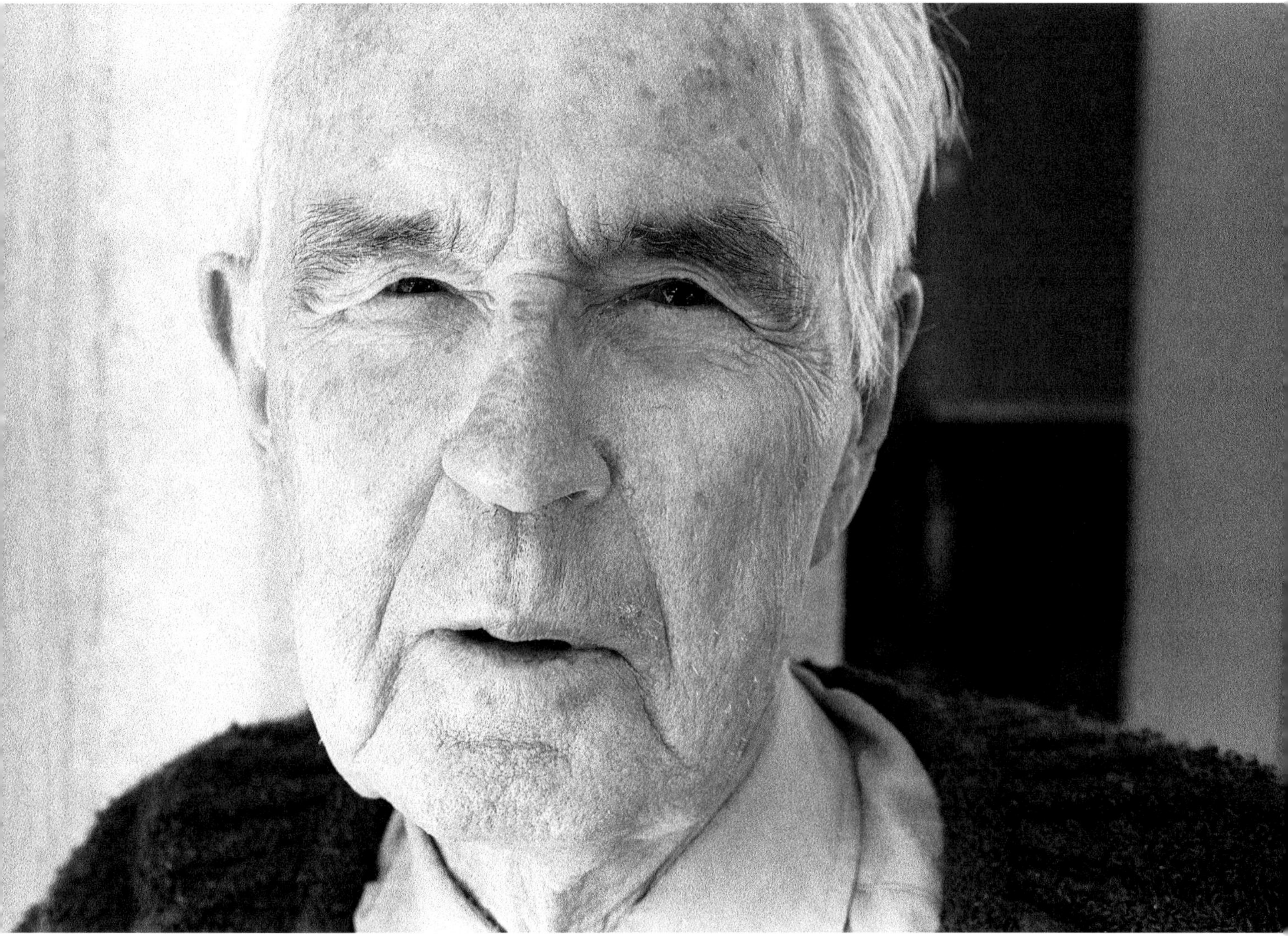

Robert B. Laughlin *1950

The Nobel Prize in Physics 1998 was awarded jointly to Robert B. Laughlin, Horst L. Störmer and Daniel C. Tsui "for their discovery of a new form of quantum fluid with fractionally charged excitations."

Ein große
Termin
Zeitraum
Ausflugsziel
Täglich
Freitag -
Sonntag
Sonntag

Leon M. Lederman *1922 †2018

The Nobel Prize in Physics 1988 was awarded jointly to Leon M. Lederman, Melvin Schwartz and Jack Steinberger "for the neutrino beam method and the demonstration of the doublet structure of the leptons through the discovery of the muon neutrino."

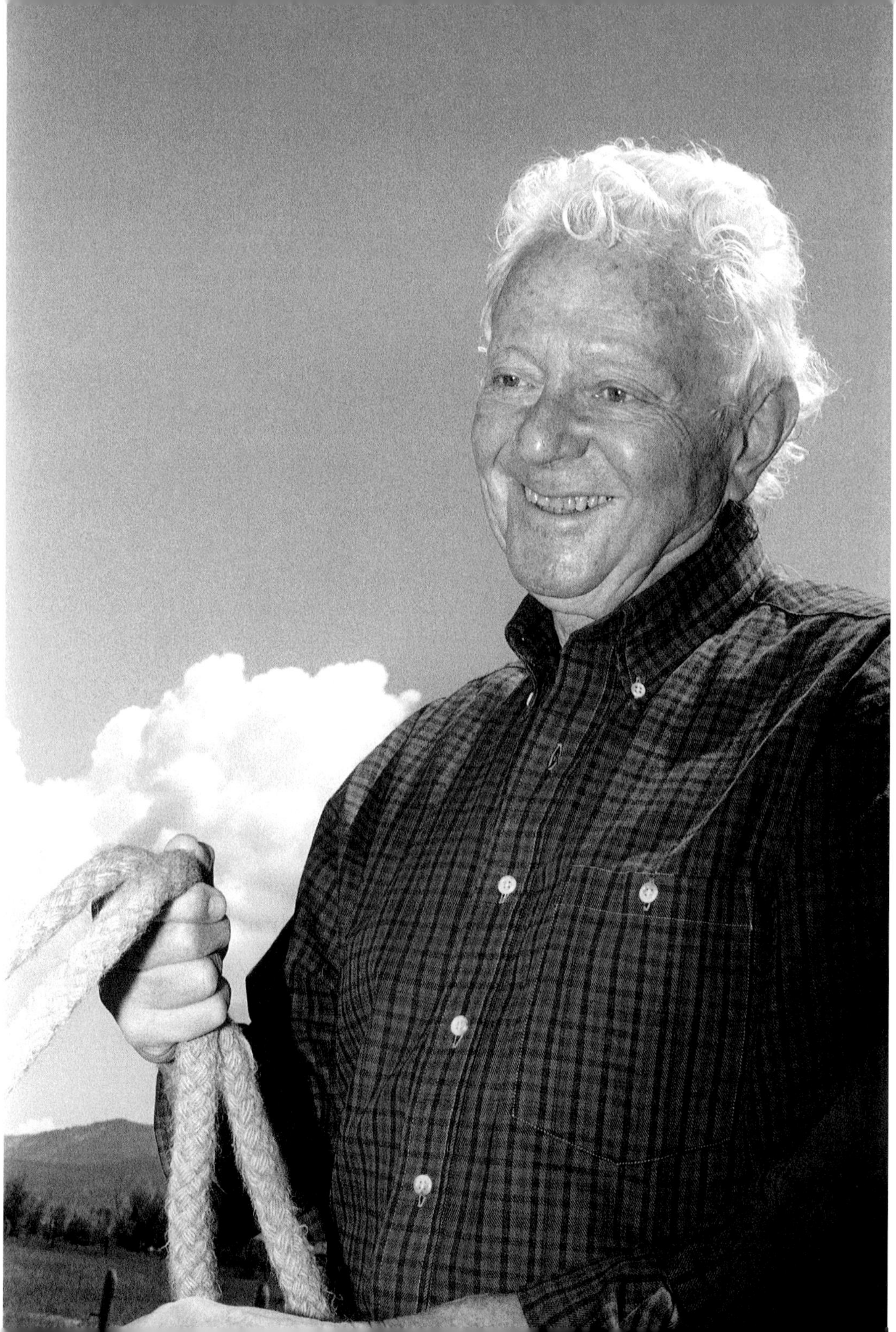

David M. Lee *1931

The Nobel Prize in Physics 1996 was awarded jointly to David M. Lee, Douglas D. Osheroff and Robert C. Richardson "for their discovery of superfluidity in helium-3."

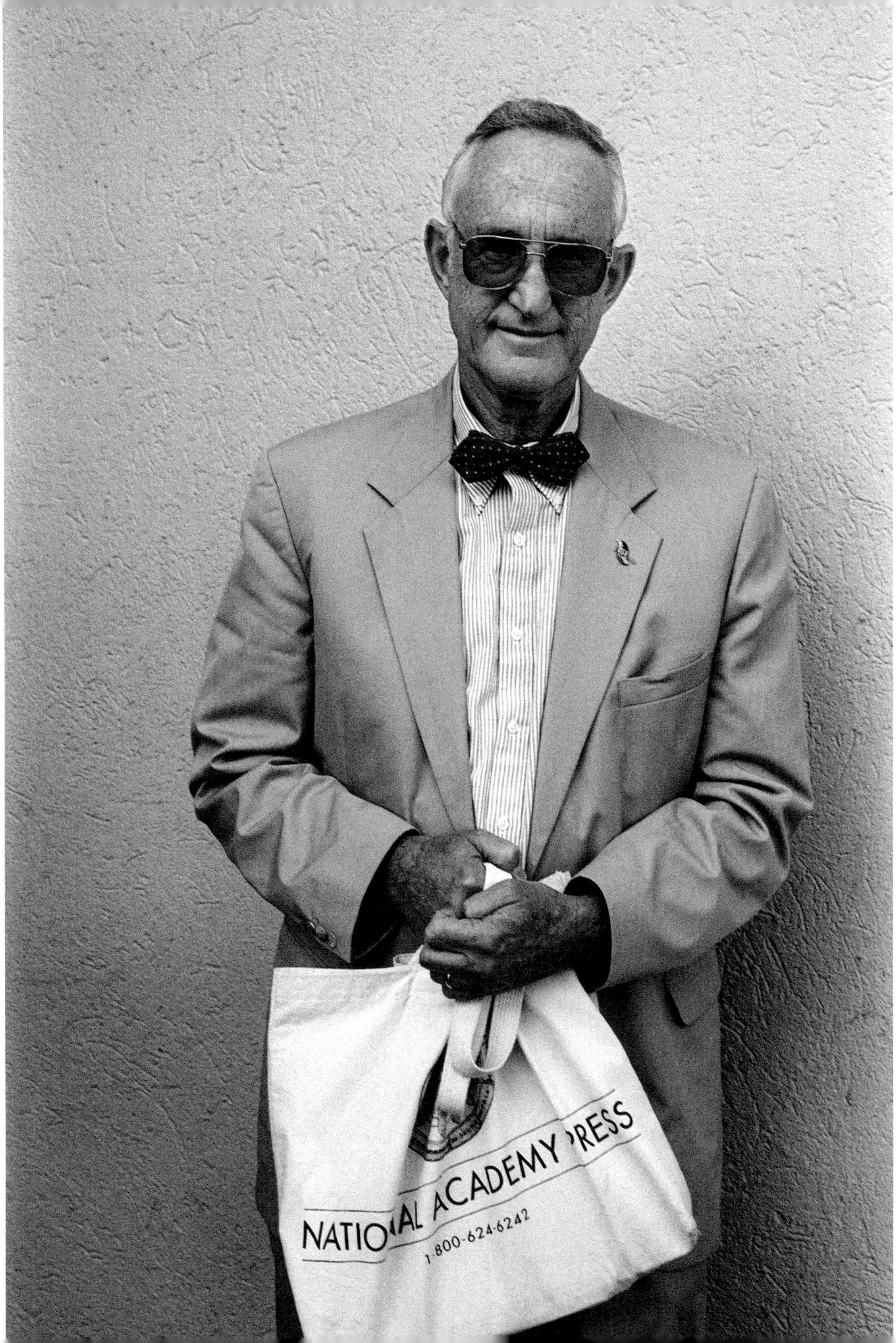
NATIO AL ACADEMY PRESS
1-800-624-6242

Tsung-Dao (T. D.) Lee *1926

The Nobel Prize in Physics 1957 was awarded jointly to Chen Ning Yang and Tsung-Dao (T. D.) Lee "for their penetrating investigation of the so-called parity laws which has led to important discoveries regarding the elementary particles."

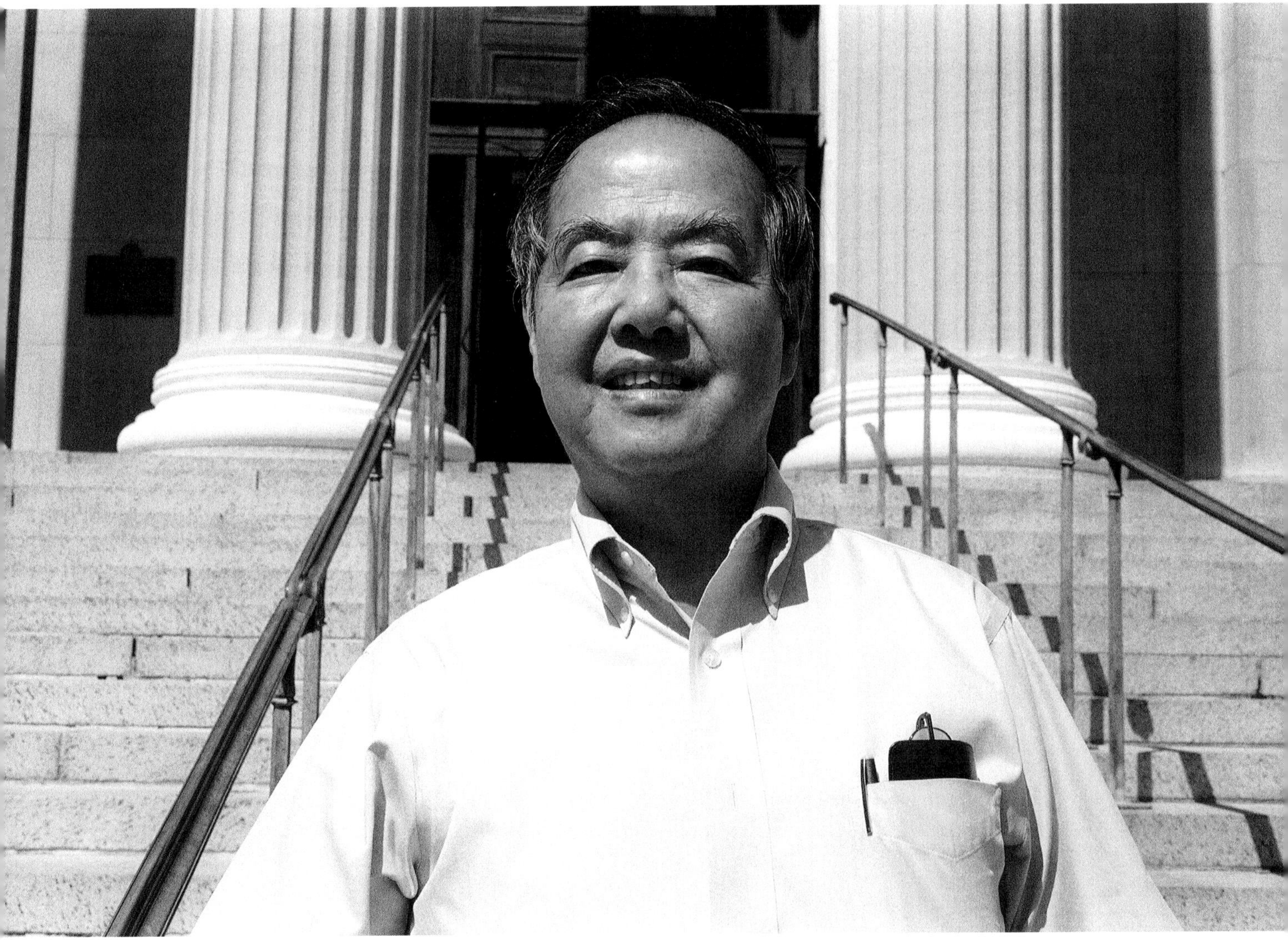

Anthony J. Leggett *1938

The Nobel Prize in Physics 2003 was awarded jointly to Alexei A. Abrikosov, Vitaly L. Ginzburg and Anthony J. Leggett "for pioneering contributions to the theory of superconductors and superfluids."

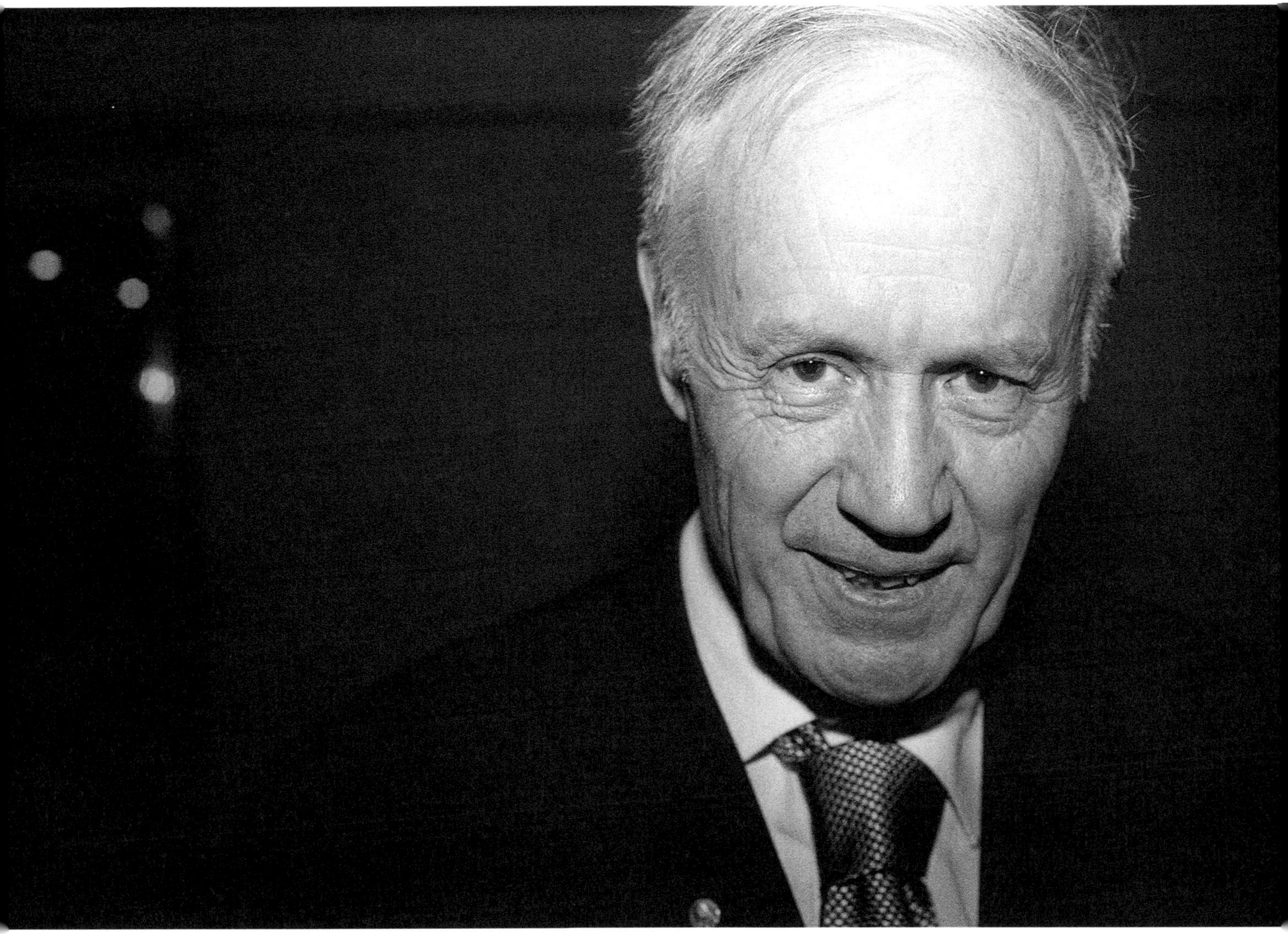

Toshihide Maskawa *1940

The Nobel Prize in Physics 2008 was divided, one half awarded to Yoichiro Nambu "for the discovery of the mechanism of spontaneous broken symmetry in subatomic physics", the other half jointly to Makoto Kobayashi and Toshihide Maskawa "for the discovery of the origin of the broken symmetry which predicts the existence of at least three families of quarks in nature."

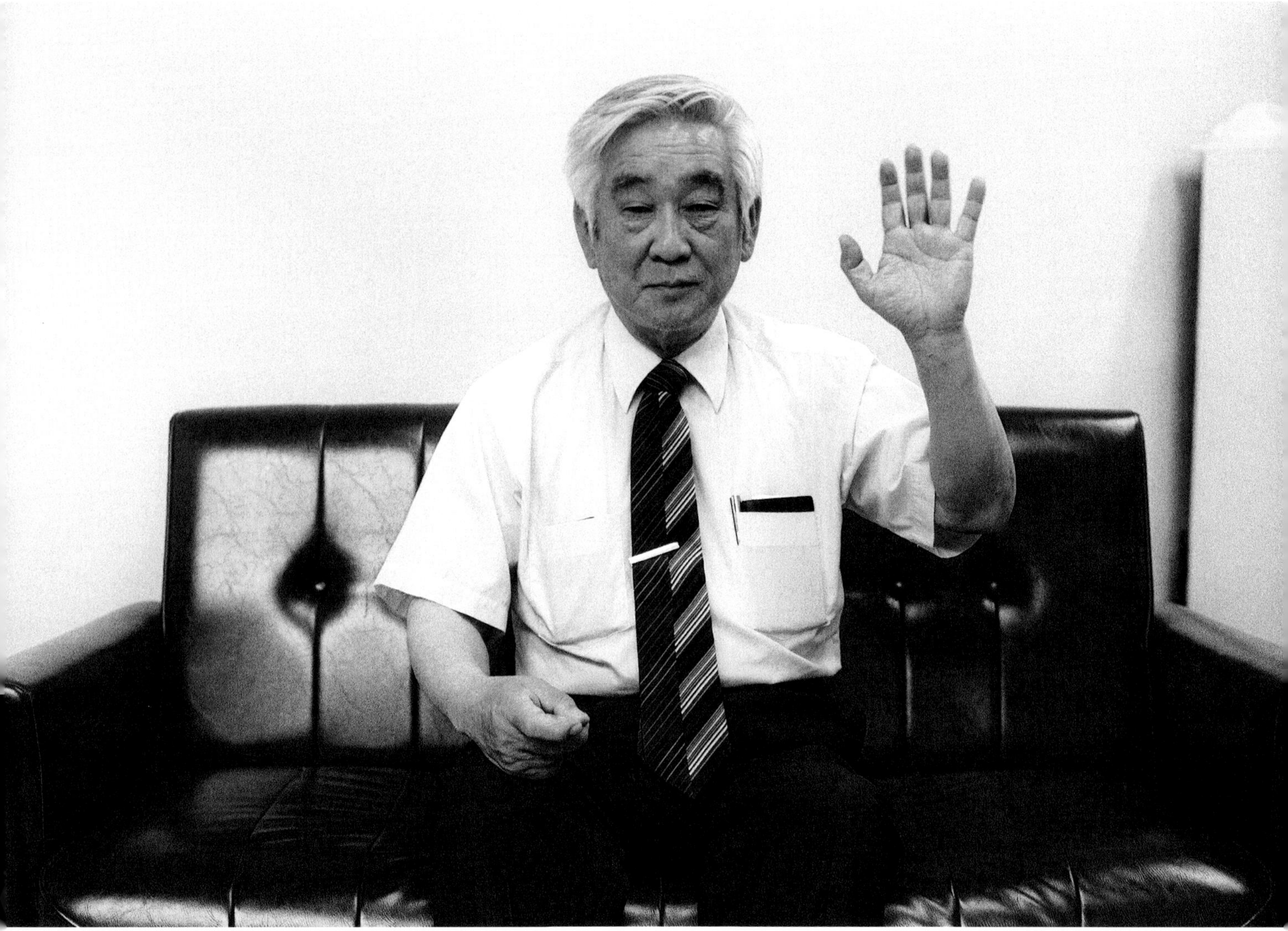

John C. Mather *1946

The Nobel Prize in Physics 2006 was awarded jointly to John C. Mather and George F. Smoot "for their discovery of the blackbody form and anisotropy of the cosmic microwave background radiation."

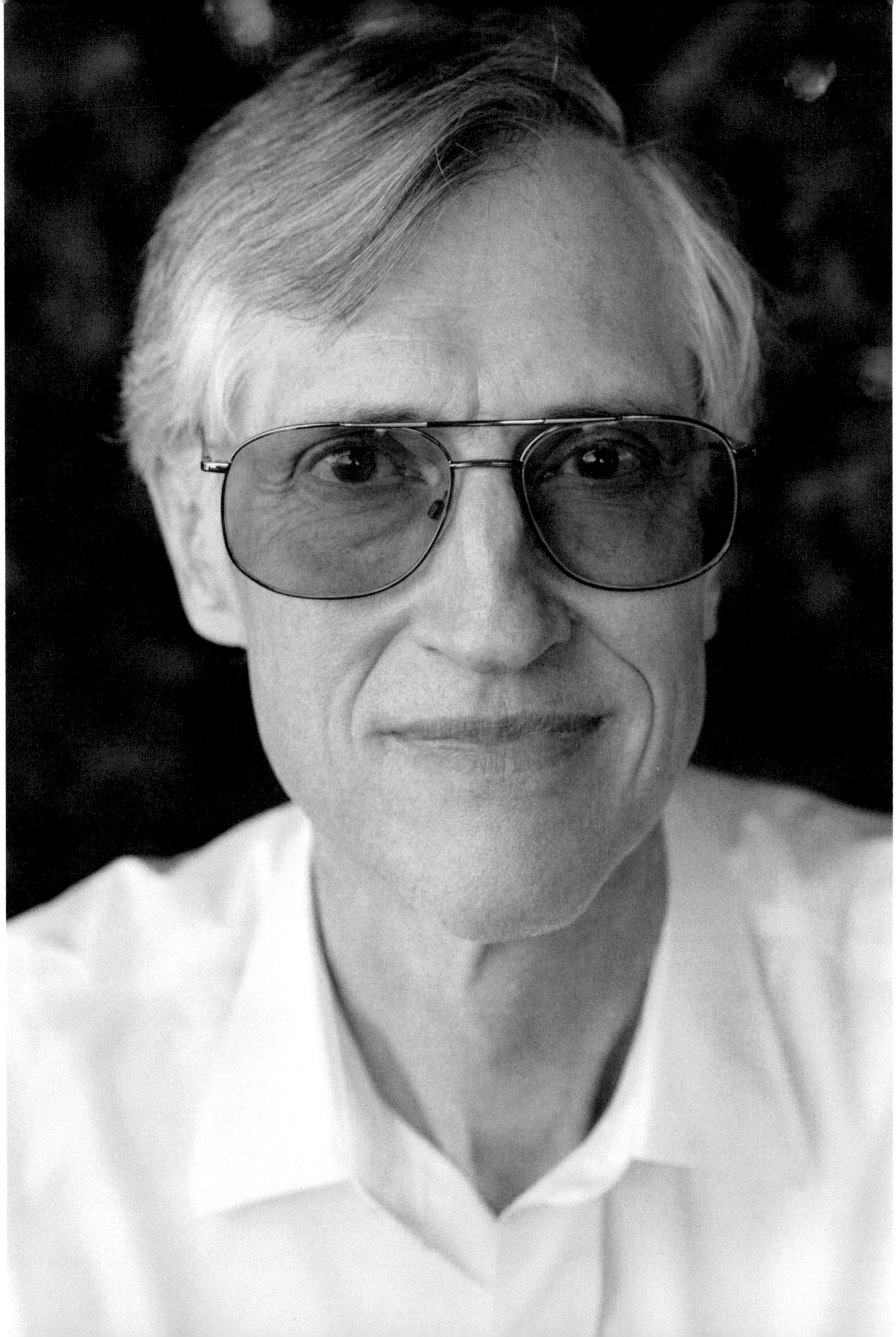

Arthur B. McDonald *1943

The Nobel Prize in Physics 2015 was awarded jointly to Takaaki Kajita and Arthur B. McDonald "for the discovery of neutrino oscillations, which shows that neutrinos have mass."

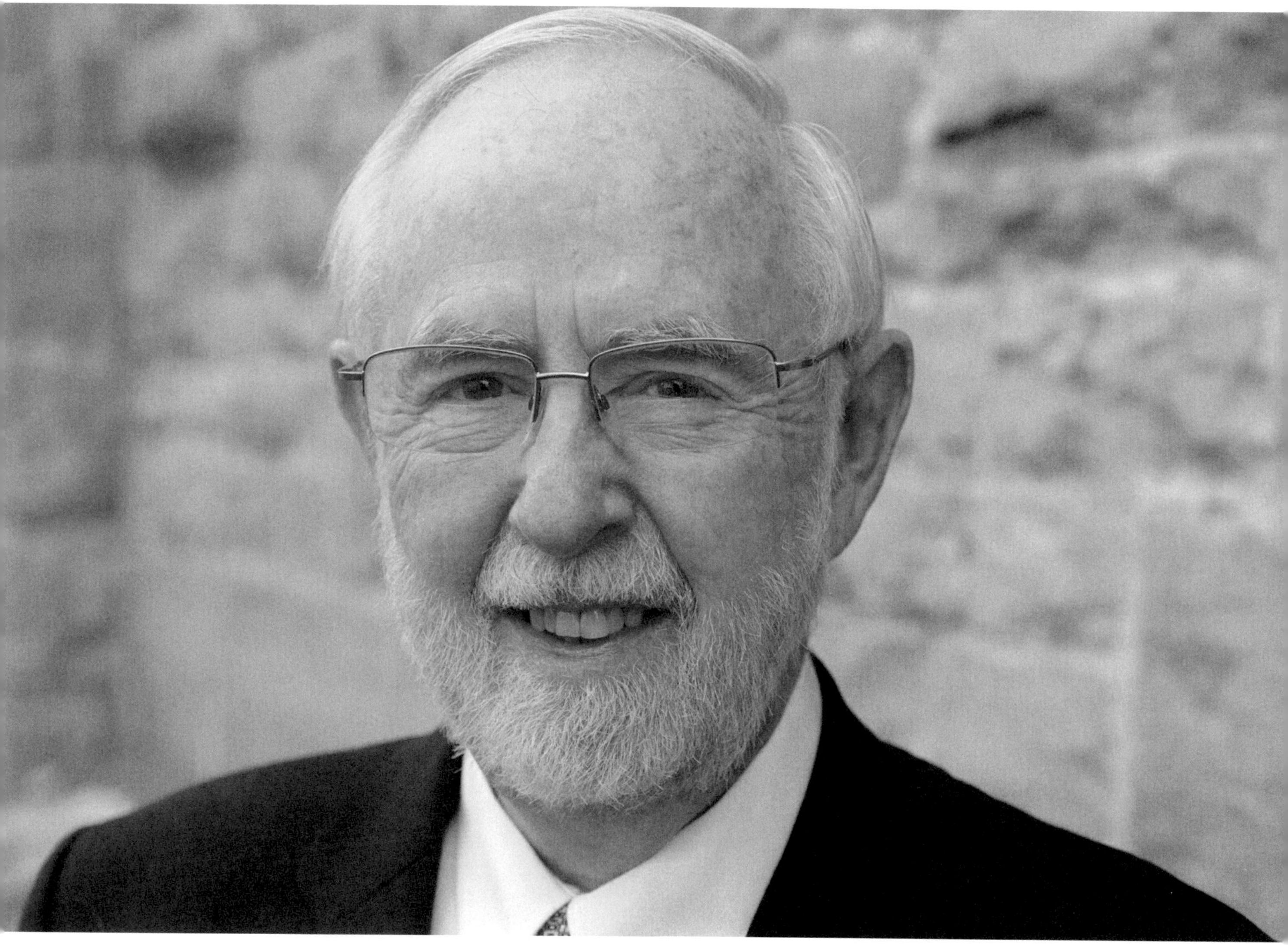

Simon van der Meer *1925 †2011

The Nobel Prize in Physics 1984 was awarded jointly to Carlo Rubbia and Simon van der Meer "for their decisive contributions to the large project, which led to the discovery of the field particles W and Z, communicators of weak interaction."

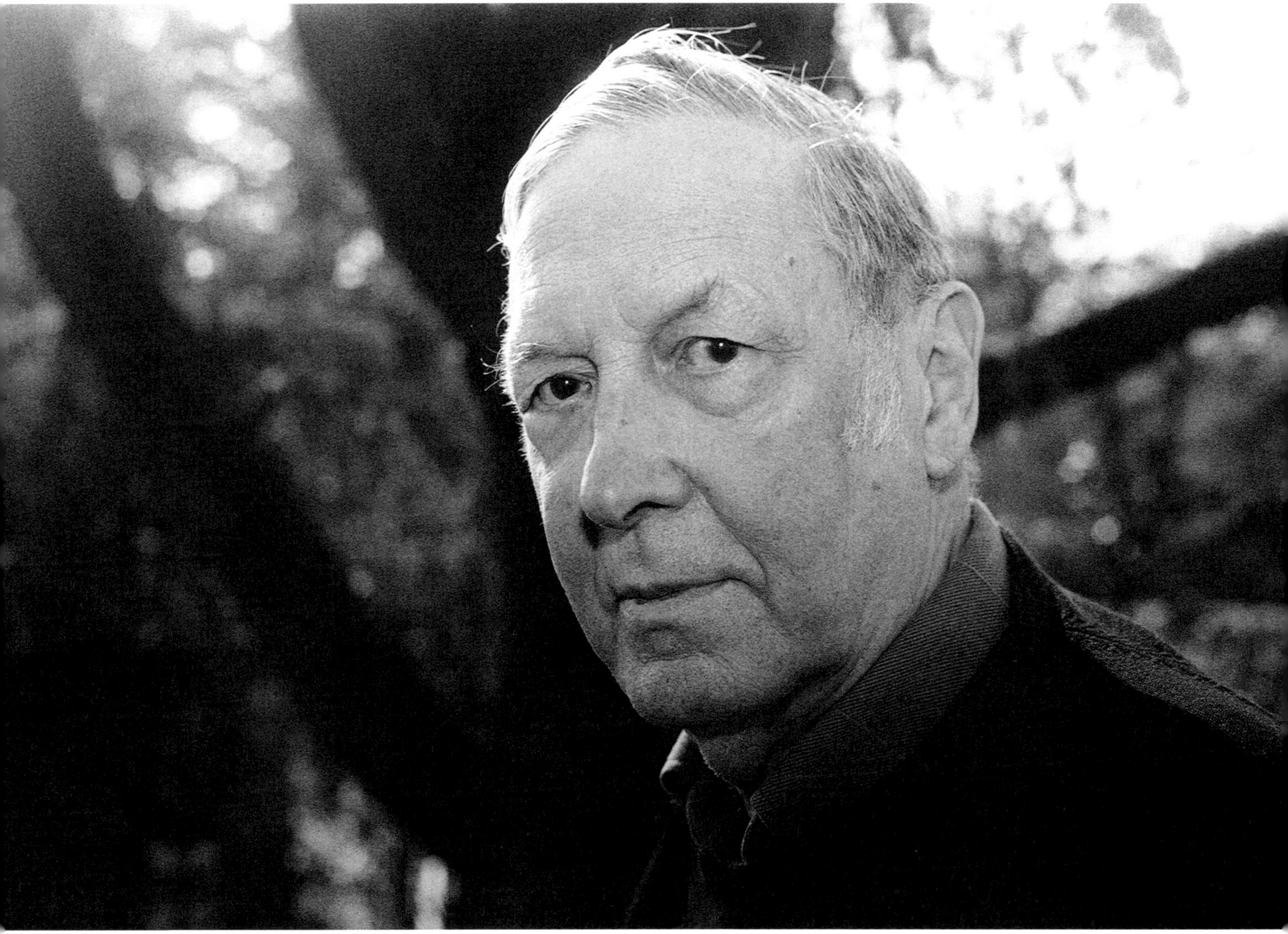

Rudolf Ludwig Mössbauer *1929 †2011

The Nobel Prize in Physics 1961 was divided equally between Robert Hofstadter "for his pioneering studies of electron scattering in atomic nuclei and for his thereby achieved discoveries concerning the structure of the nucleons" and Rudolf Ludwig Mössbauer "for his researches concerning the resonance absorption of gamma radiation and his discovery in this connection of the effect which bears his name."

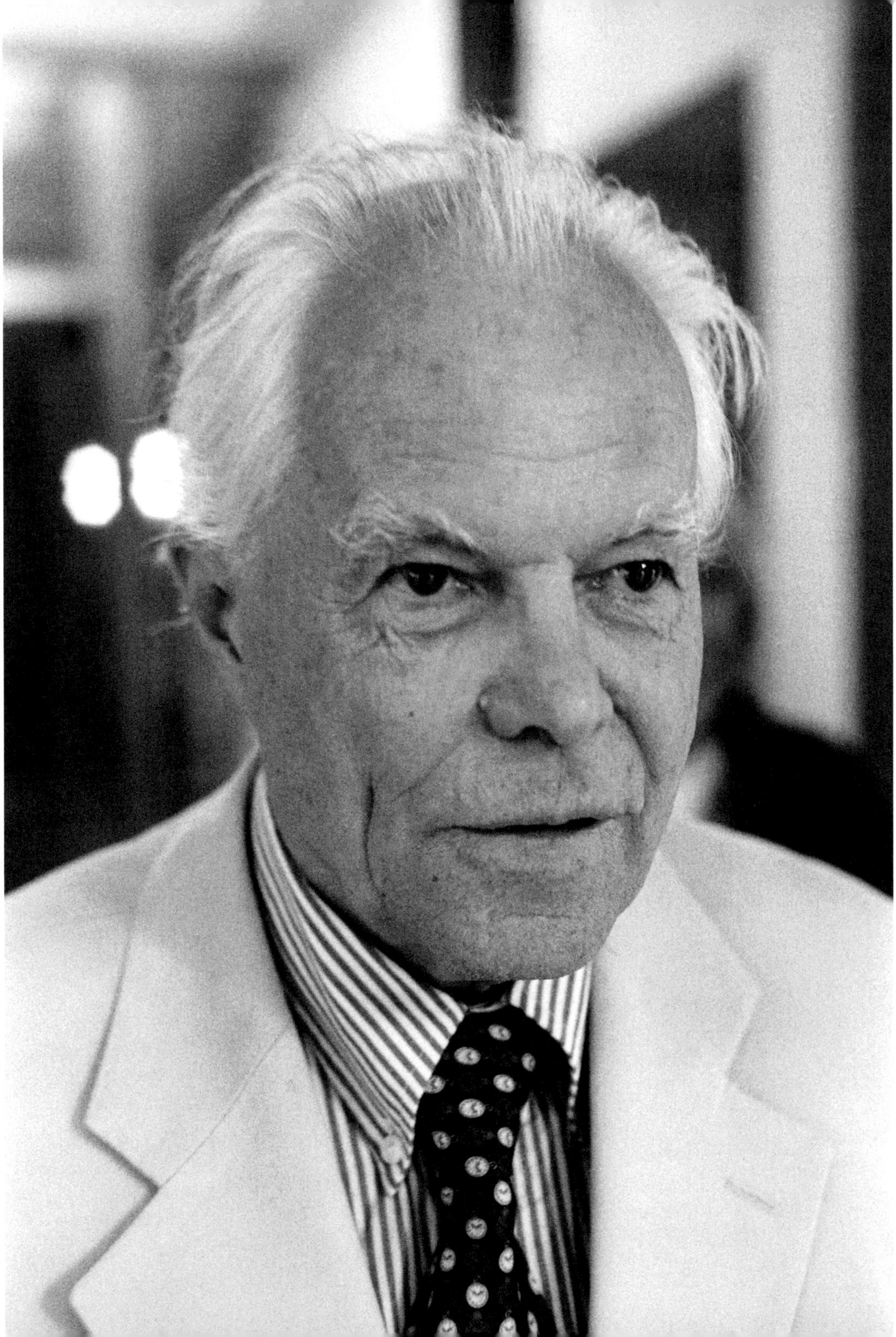

Ben Roy Mottelson *1938

The Nobel Prize in Physics 1975 was awarded jointly to Aage Niels Bohr, Ben Roy Mottelson and Leo James Rainwater "for the discovery of the connection between collective motion and particle motion in atomic nuclei and the development of the theory of the structure of the atomic nucleus based on this connection."

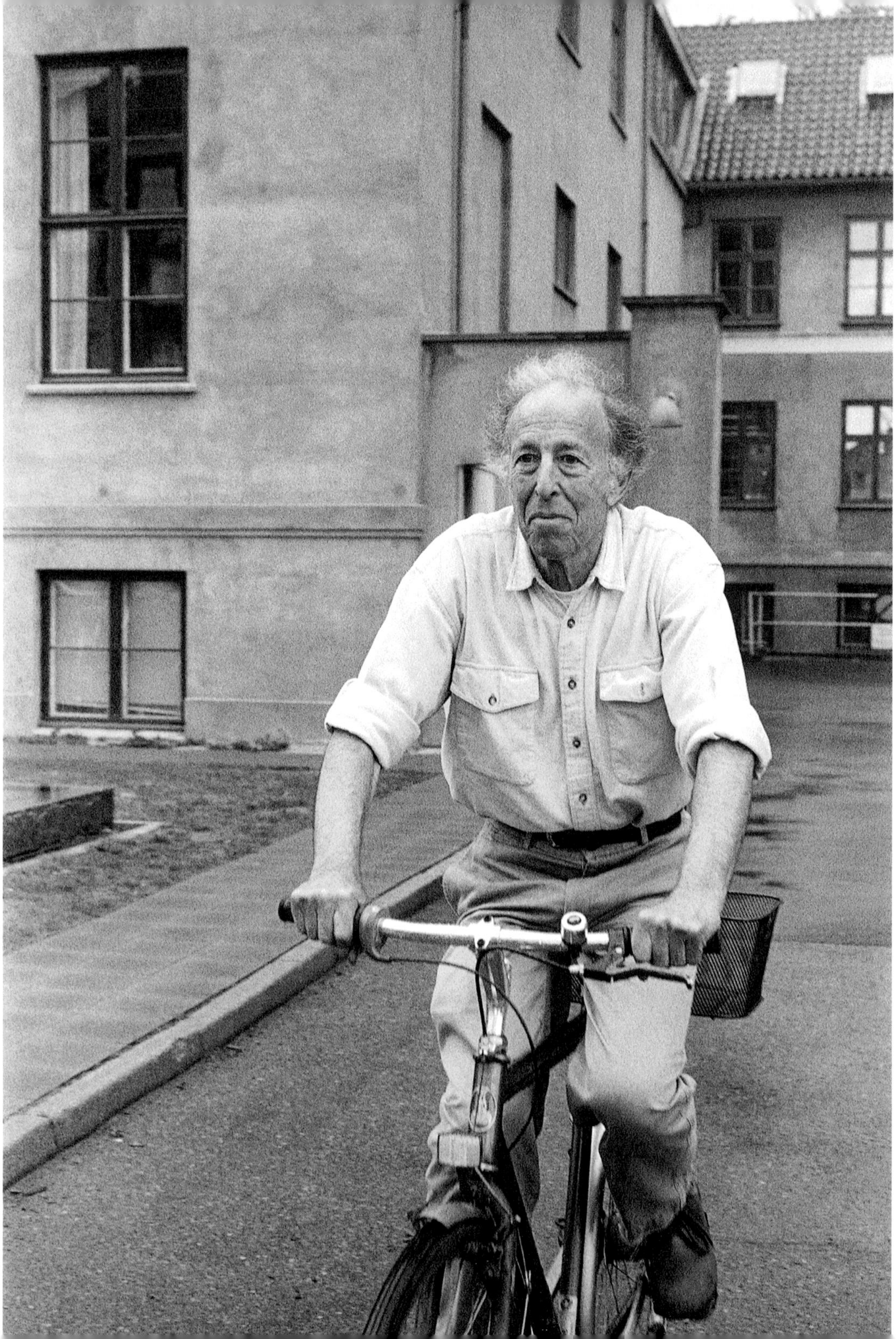

Gérard Mourou *1944

The Nobel Prize in Physics 2018 was awarded "for groundbreaking inventions in the field of laser physics" with one half to Arthur Ashkin "for the optical tweezers and their application to biological systems", the other half jointly to Gérard Mourou and Donna Strickland "for their method of generating high-intensity, ultra-short optical pulses."

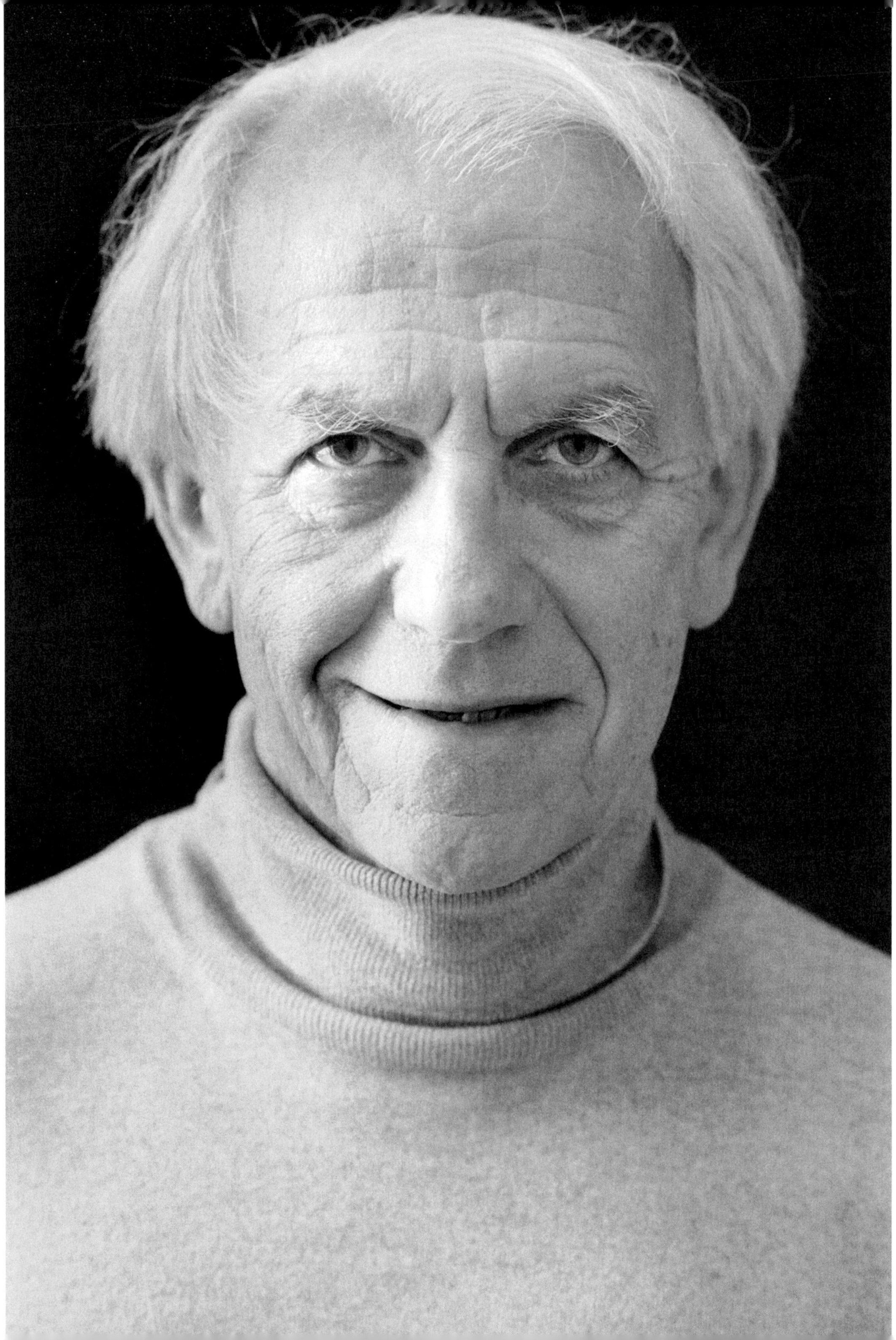

K. Alexander Müller *1927

The Nobel Prize in Physics 1987 was awarded jointly to J. Georg Bednorz and K. Alexander Müller "for their important break-through in the discovery of superconductivity in ceramic materials."

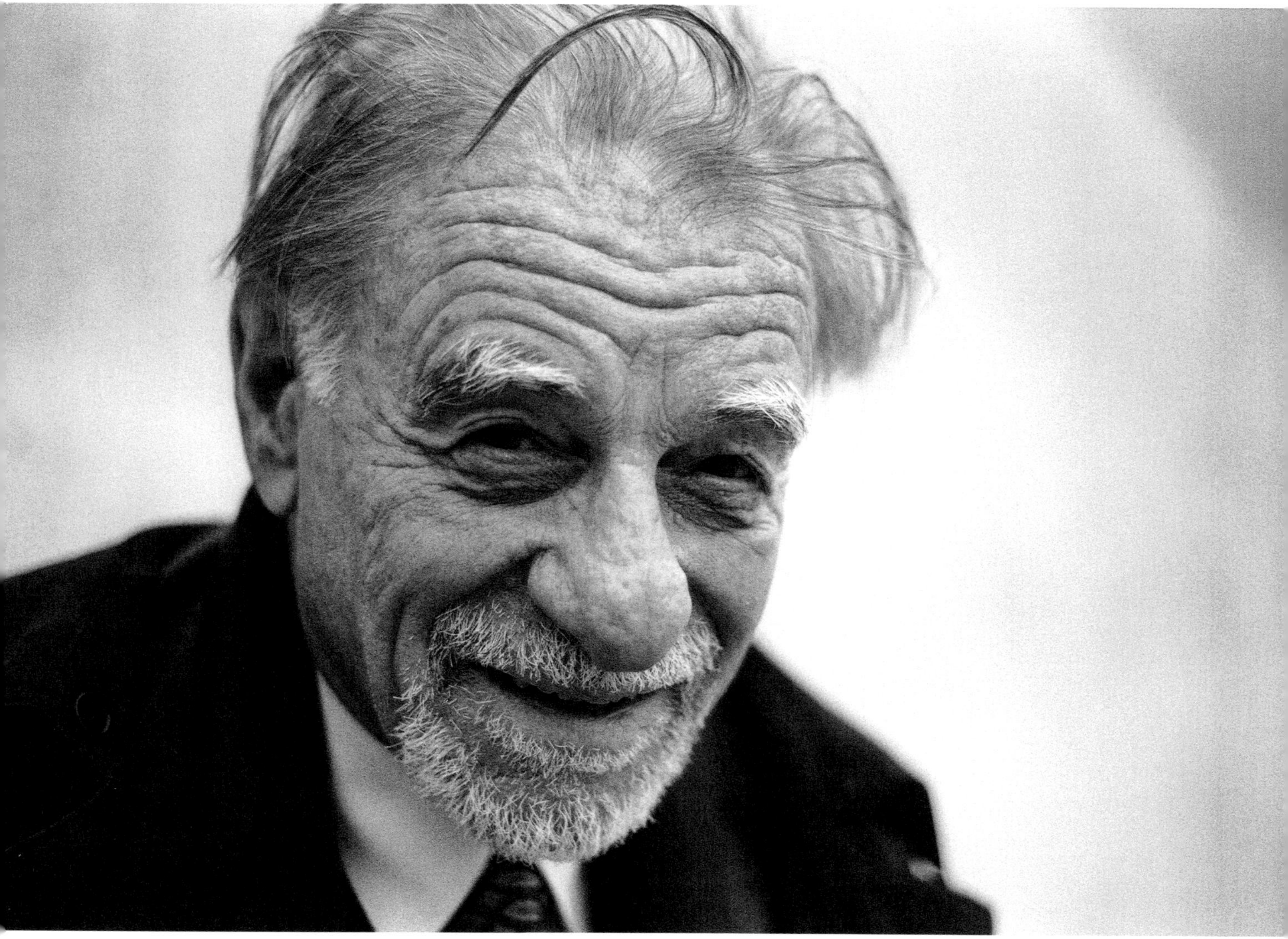

Shuji Nakamura *1954

The Nobel Prize in Physics 2014 was awarded jointly to Isamu Akasaki, Hiroshi Amano and Shuji Nakamura "for the invention of efficient blue light-emitting diodes which has enabled bright and energy-saving white light sources."

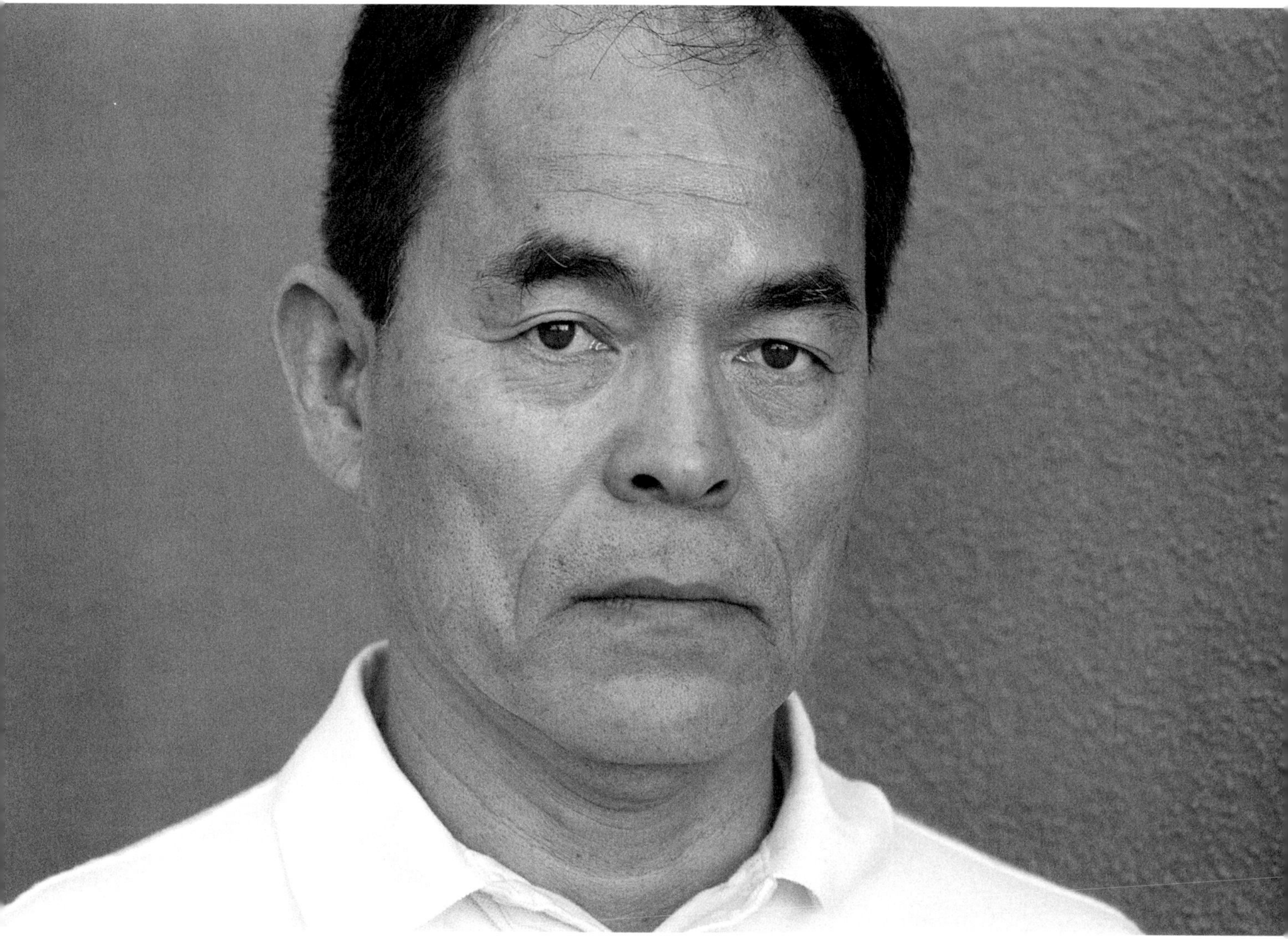

Yoichiro Nambu *1921 †2015

The Nobel Prize in Physics 2008 was divided, one half awarded to Yoichiro Nambu "for the discovery of the mechanism of spontaneous broken symmetry in subatomic physics", the other half jointly to Makoto Kobayashi and Toshihide Maskawa "for the discovery of the origin of the broken symmetry which predicts the existence of at least three families of quarks in nature."

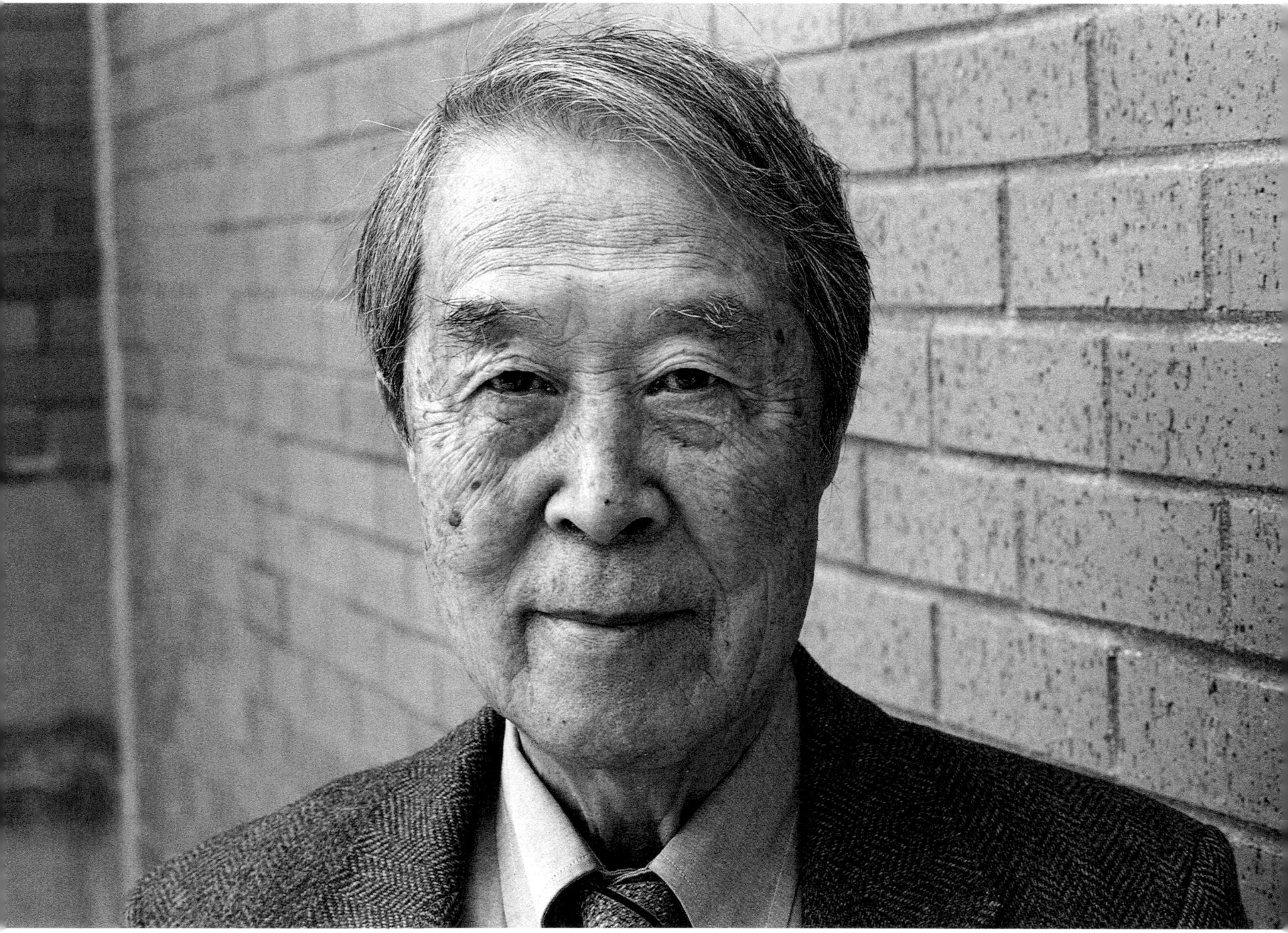

Konstantin Novoselov *1974

The Nobel Prize in Physics 2010 was awarded jointly to Andre Geim and Konstantin Novoselov "for groundbreaking experiments regarding the two-dimensional material graphene."

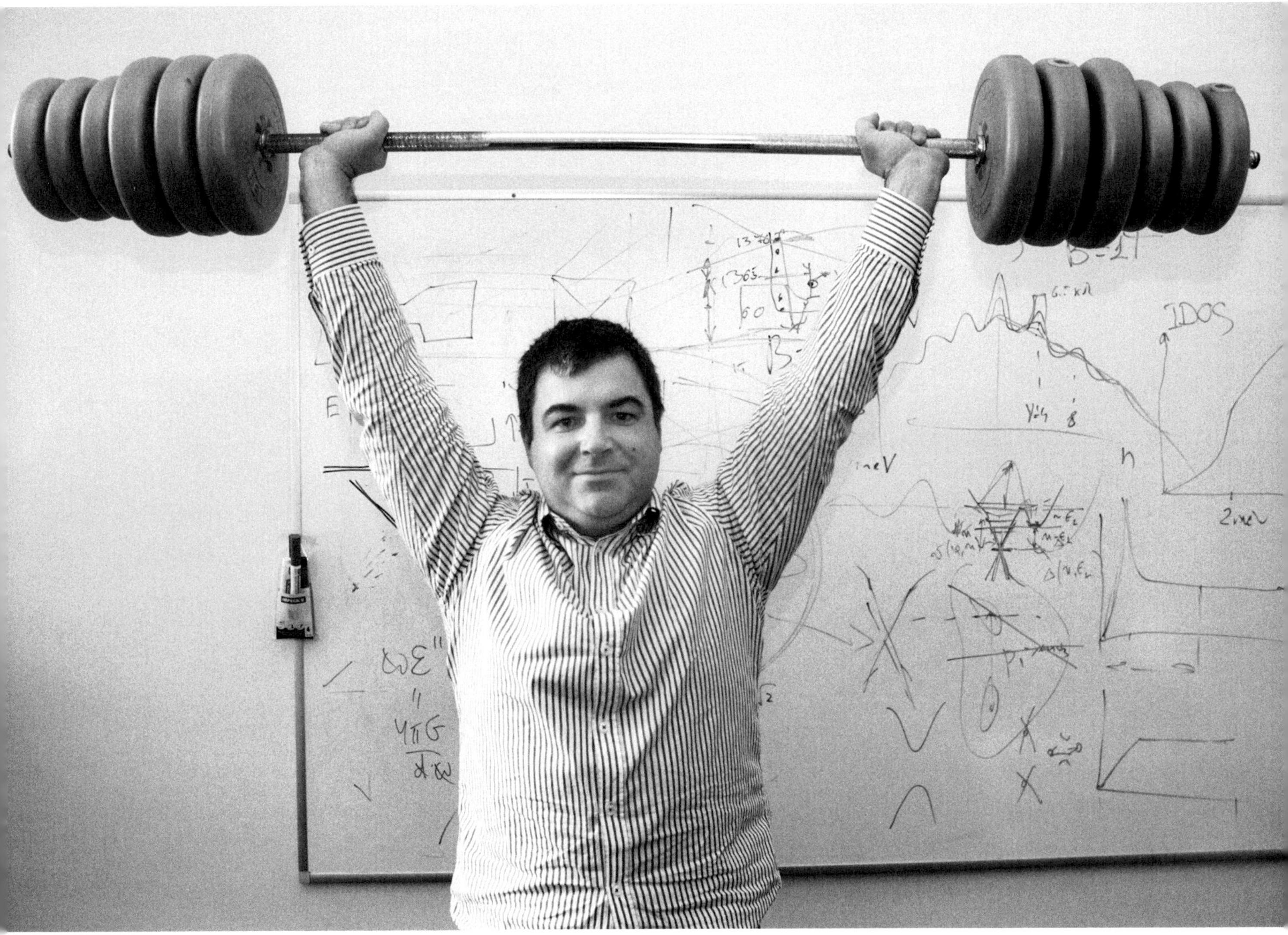
IDOS

Douglas D. Osheroff *1945

The Nobel Prize in Physics 1996 was awarded jointly to David M. Lee, Douglas D. Osheroff and Robert C. Richardson "for their discovery of superfluidity in helium-3."

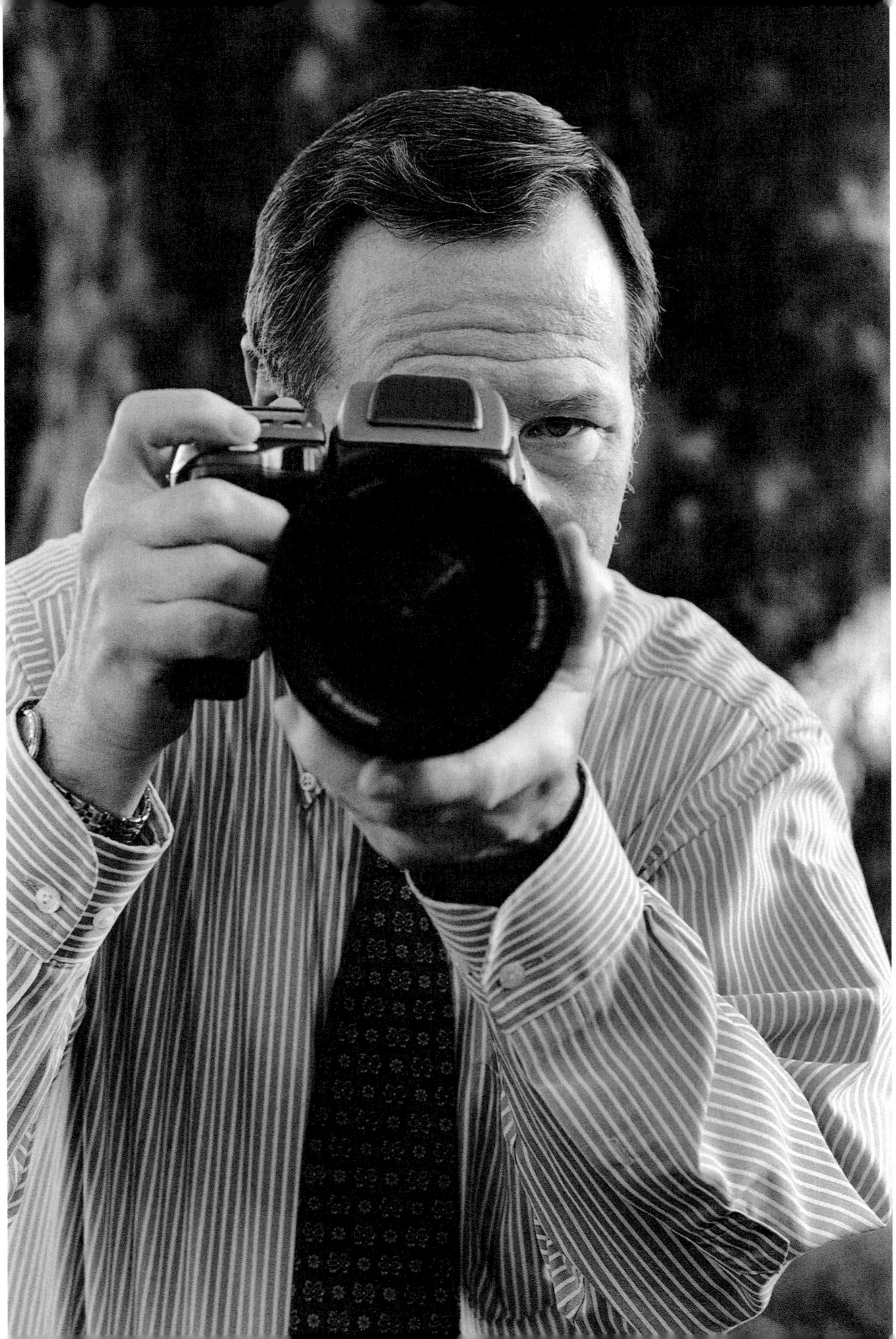

Arno Allan Penzias * 1933

The Nobel Prize in Physics 1978 was divided, one half awarded to Pyotr Leonidovich Kapitsa "for his basic inventions and discoveries in the area of low-temperature physics", the other half jointly to Arno Allan Penzias and Robert Woodrow Wilson "for their discovery of cosmic microwave background radiation."

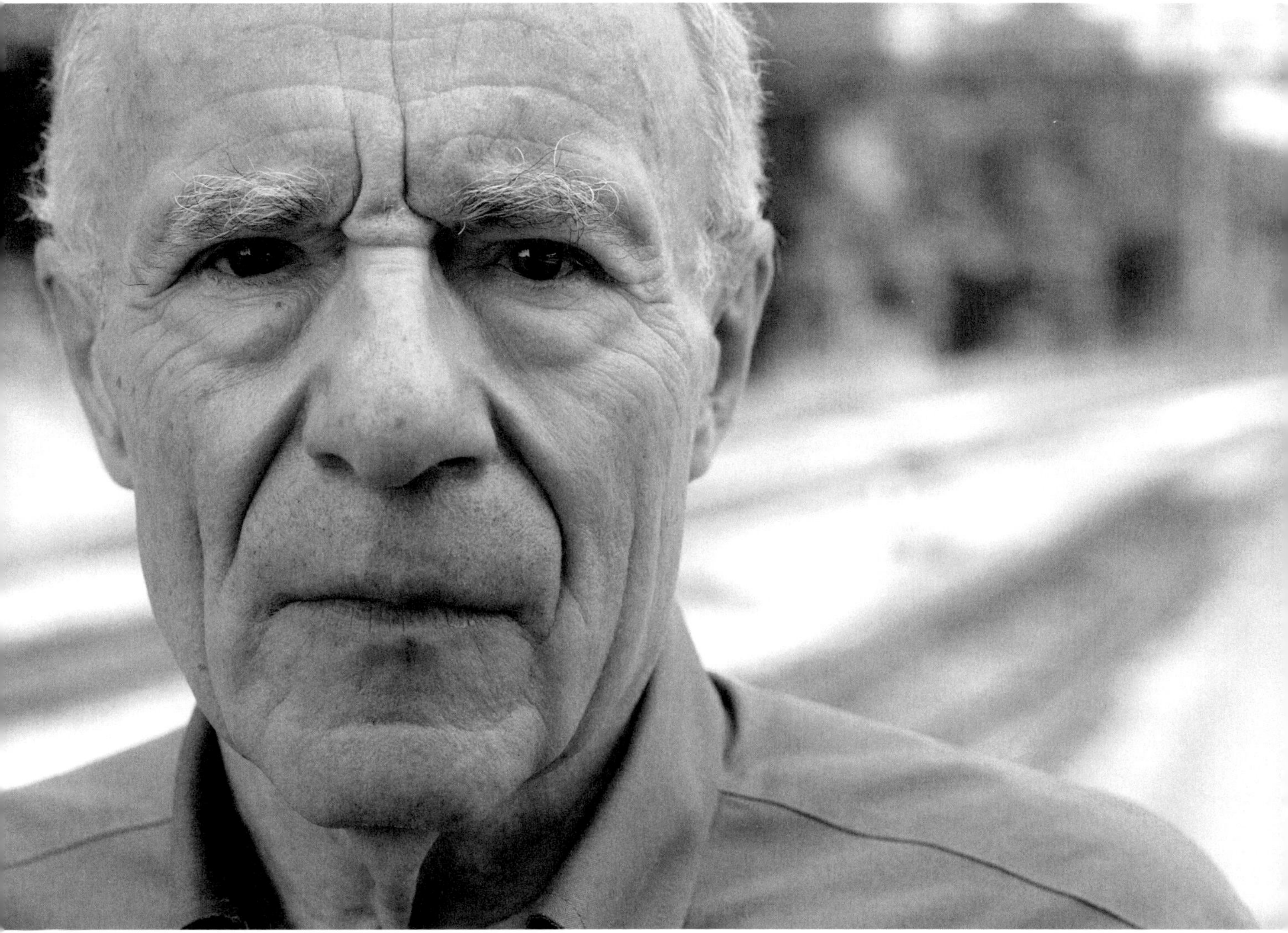

Martin L. Perl *1927 †2014

The Nobel Prize in Physics 1995 was awarded "for pioneering experimental contributions to lepton physics" jointly with one half to Martin L. Perl "for the discovery of the tau lepton" and with one half to Frederick Reines "for the detection of the neutrino."

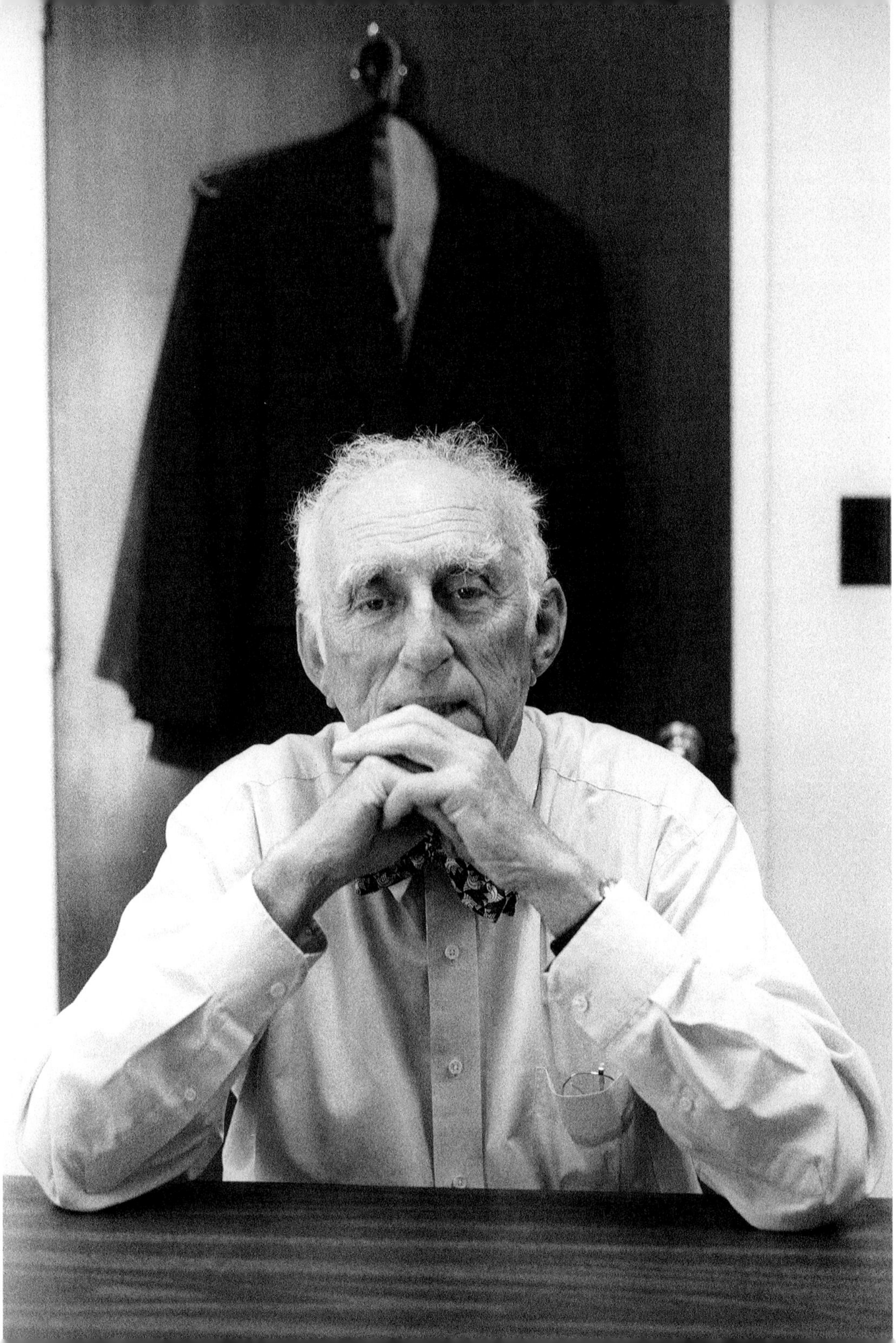

Saul Perlmutter *1959

The Nobel Prize in Physics 2011 was divided, one half awarded to Saul Perlmutter, the other half jointly to Brian P. Schmidt and Adam G. Riess "for the discovery of the accelerating expansion of the Universe through observations of distant supernovae."

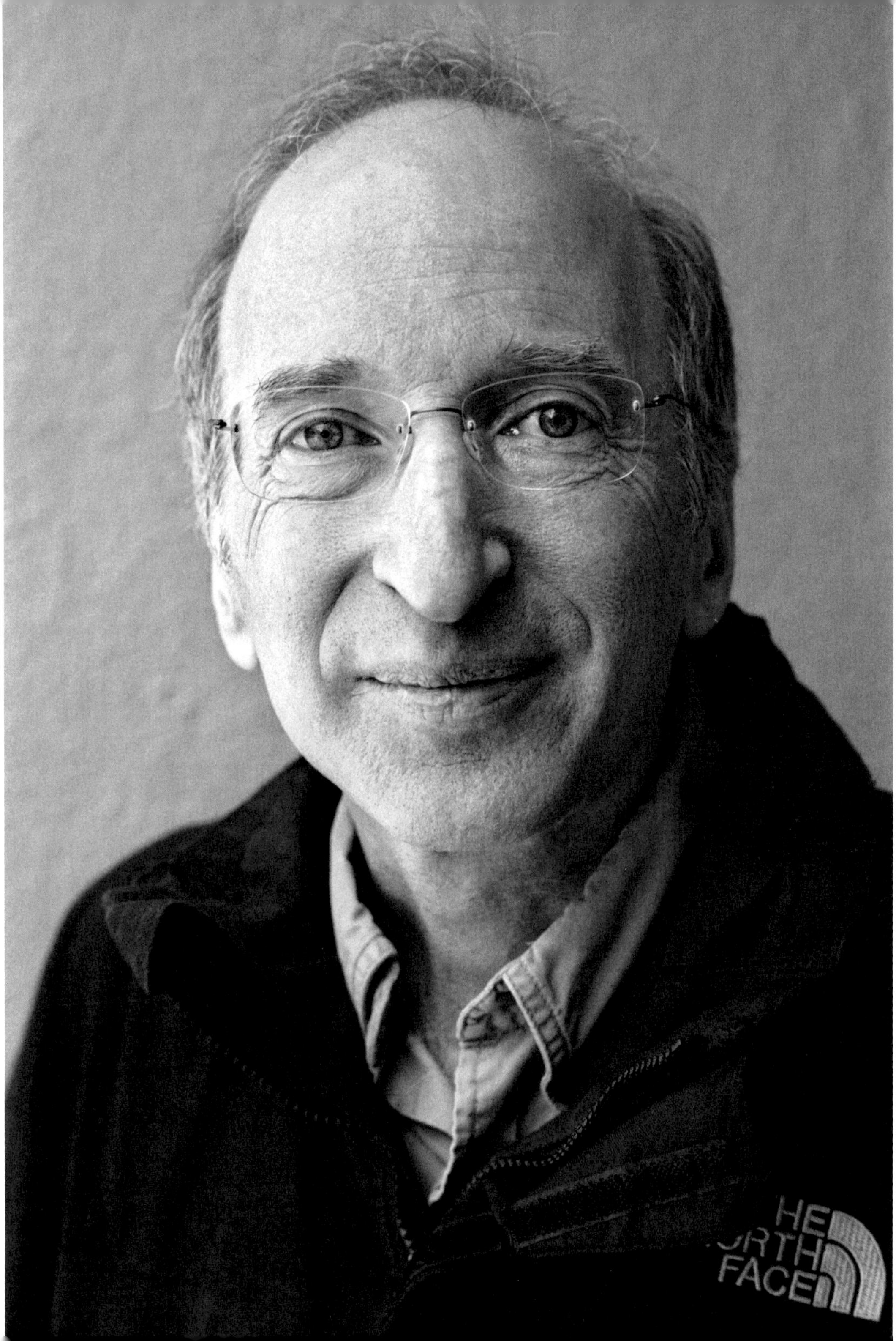

William D. Phillips *1948

The Nobel Prize in Physics 1997 was awarded jointly to Steven Chu, Claude Cohen-Tannoudji and William D. Phillips "for development of methods to cool and trap atoms with laser light."

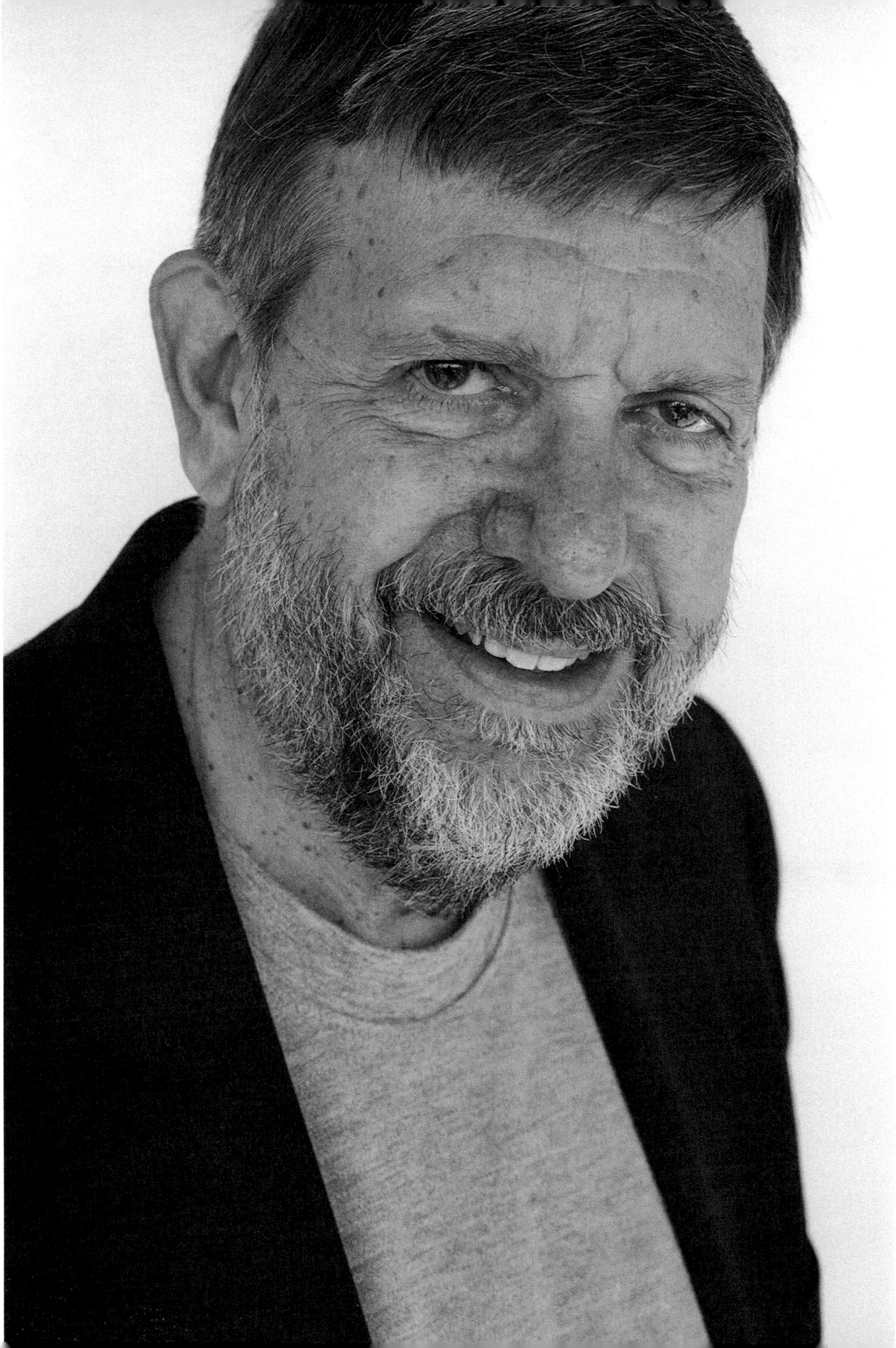

H. David Politzer *1949

The Nobel Prize in Physics 2004 was awarded jointly to David J. Gross, H. David Politzer and Frank Wilczek "for the discovery of asymptotic freedom in the theory of the strong interaction."

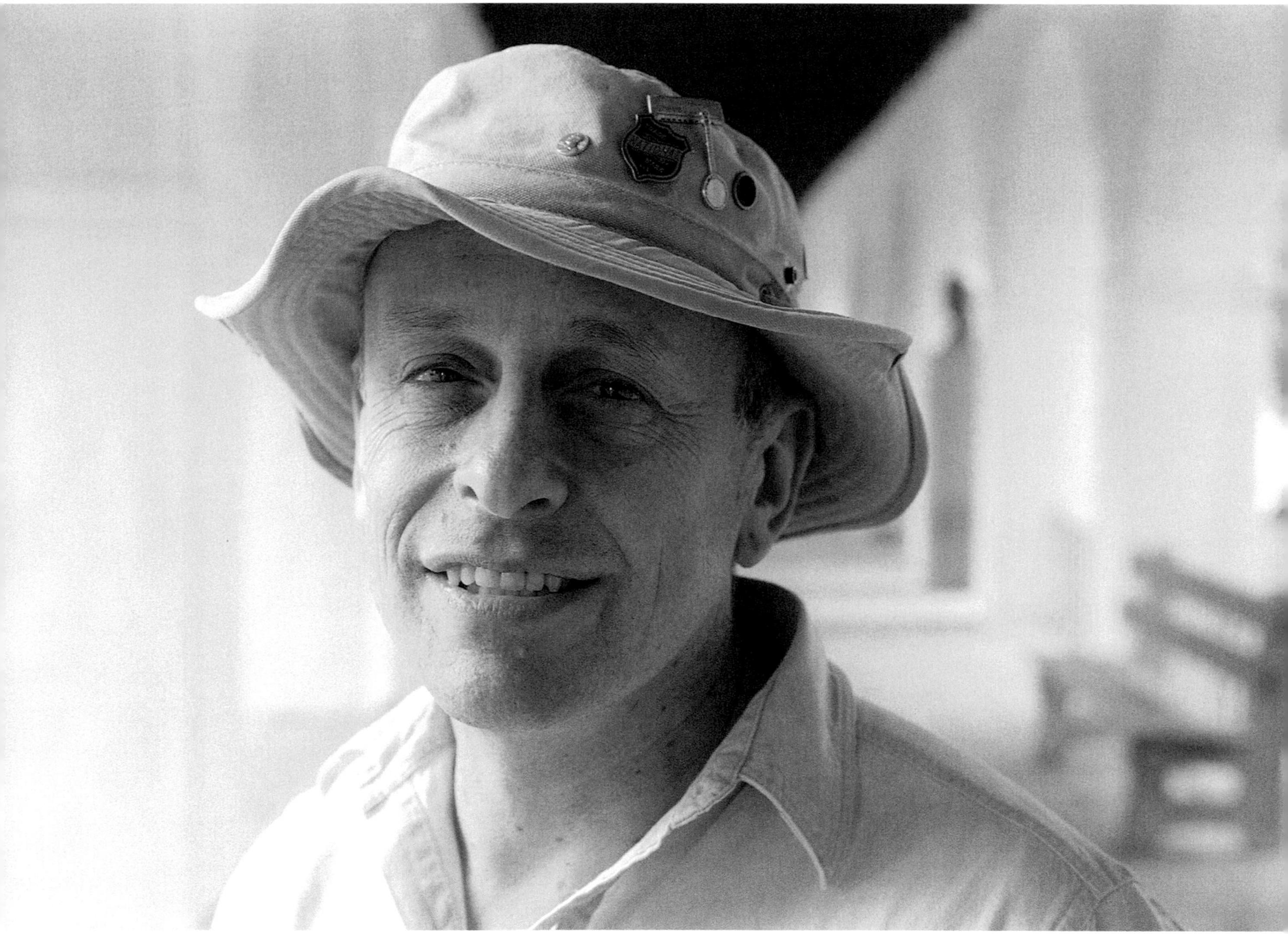

Norman F. Ramsey *1915 †2011

The Nobel Prize in Physics 1989 was divided, one half awarded to Norman F. Ramsey "for the invention of the separated oscillatory fields method and its use in the hydrogen maser and other atomic clocks", the other half jointly to Hans G. Dehmelt and Wolfgang Paul "for the development of the ion trap technique."

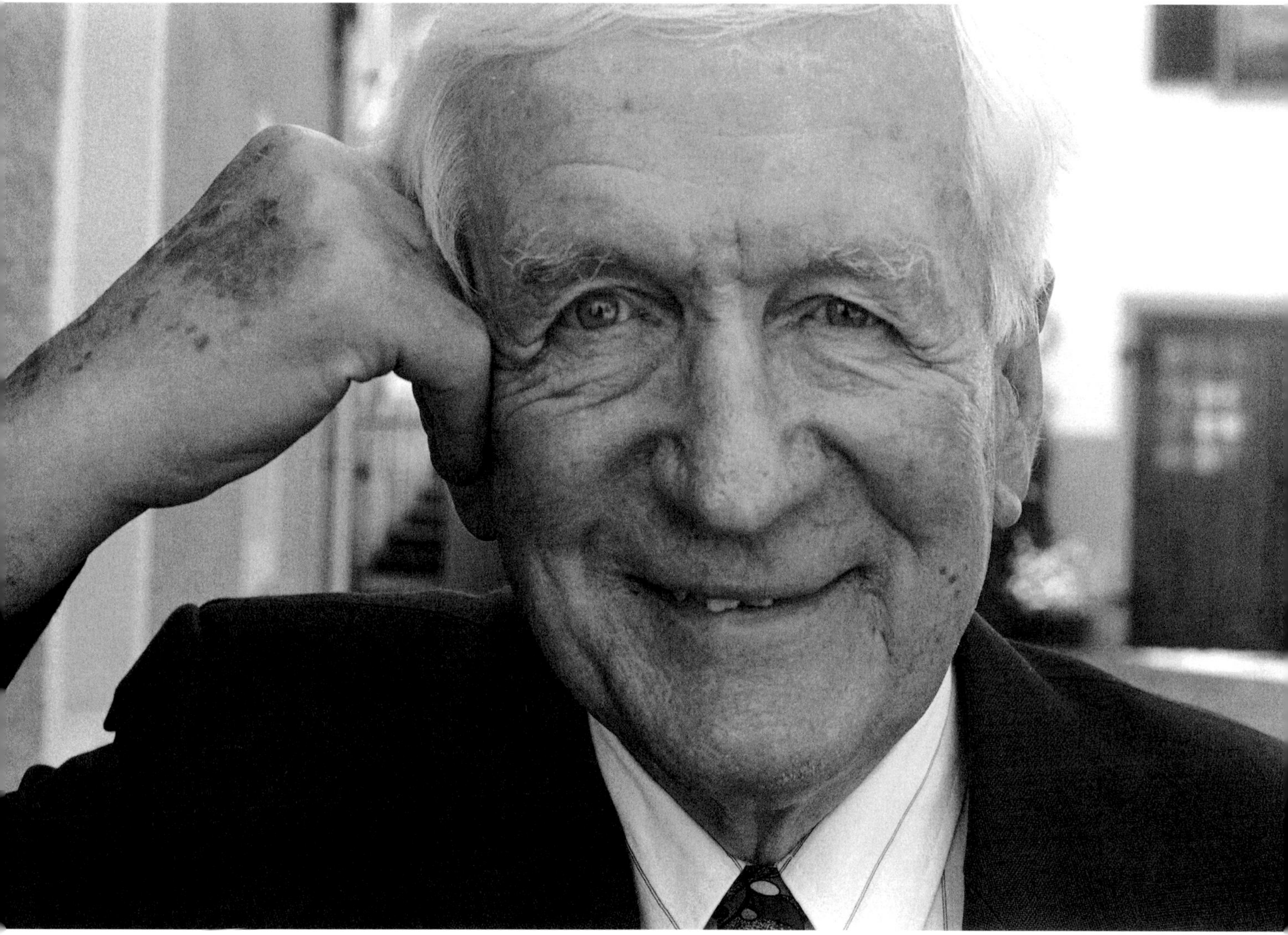

Robert C. Richardson *1937 †2013

The Nobel Prize in Physics 1996 was awarded jointly to David M. Lee, Douglas D. Osheroff and Robert C. Richardson "for their discovery of superfluidity in helium-3."

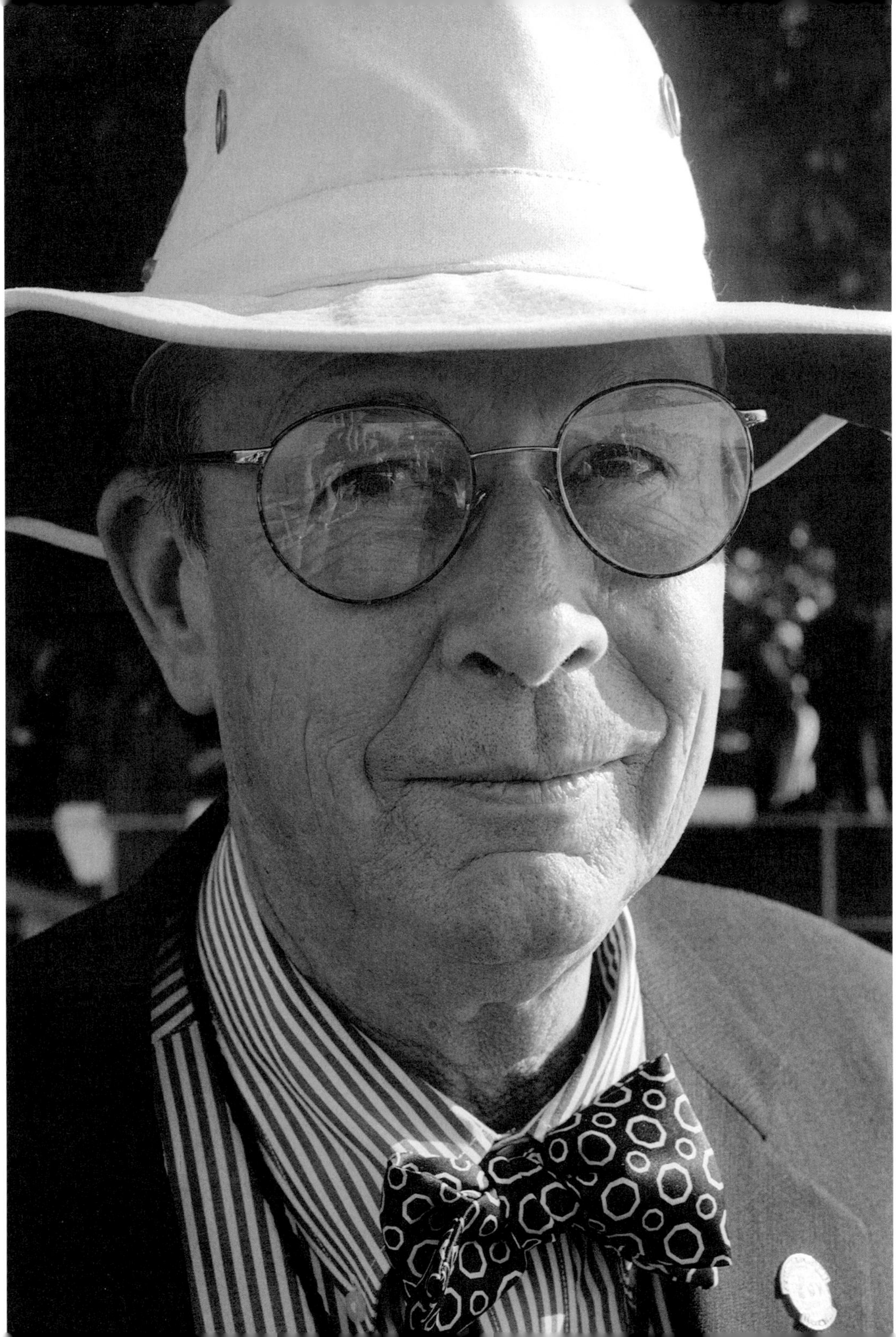

Burton Richter *1931 †2018

The Nobel Prize in Physics 1976 was awarded jointly to Burton Richter and Samuel Chao Chung Ting "for their pioneering work in the discovery of a heavy elementary particle of a new kind."

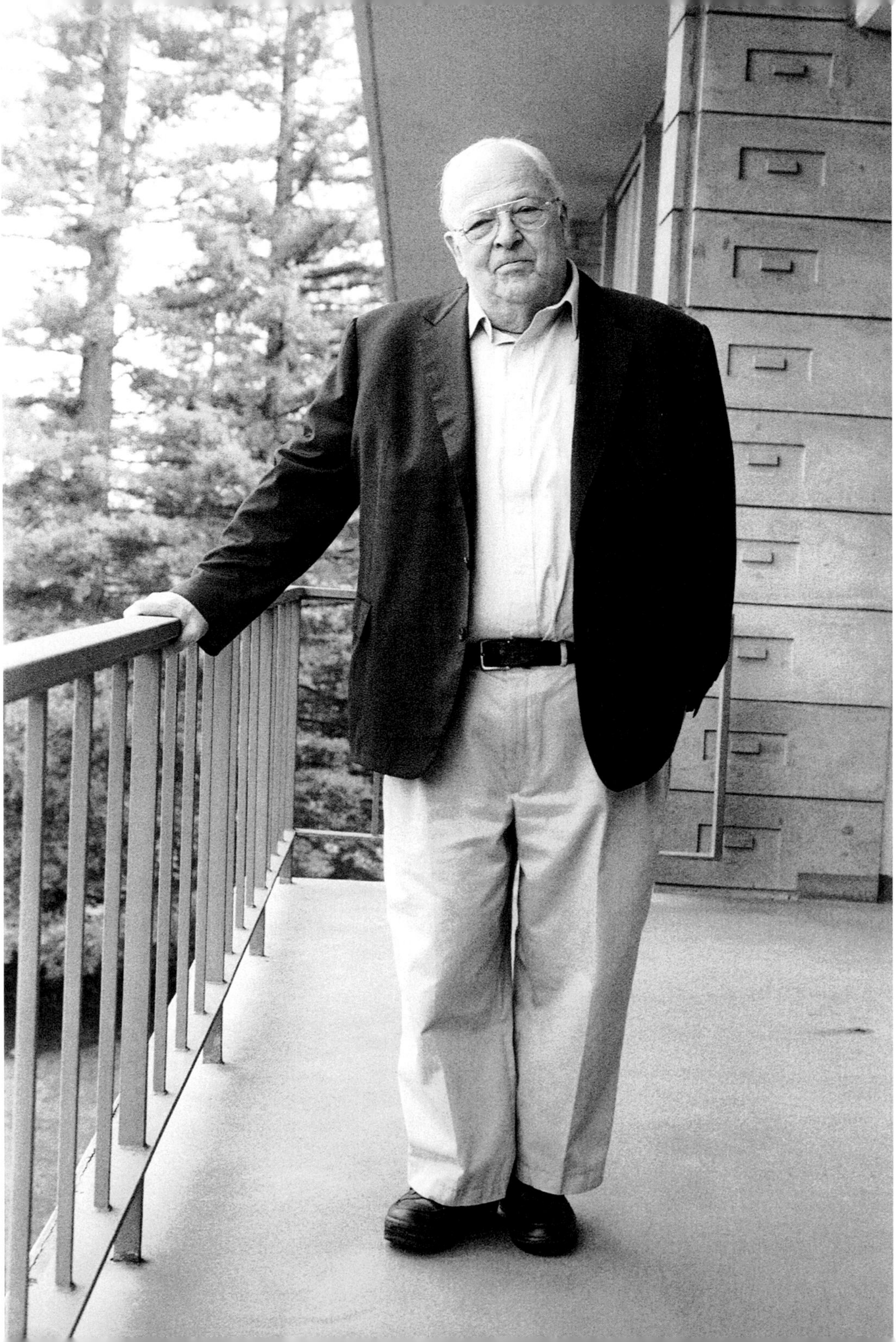

Adam G. Riess *1969

The Nobel Prize in Physics 2011 was divided, one half awarded to Saul Perlmutter, the other half jointly to Brian P. Schmidt and Adam G. Riess "for the discovery of the accelerating expansion of the Universe through observations of distant supernovae."

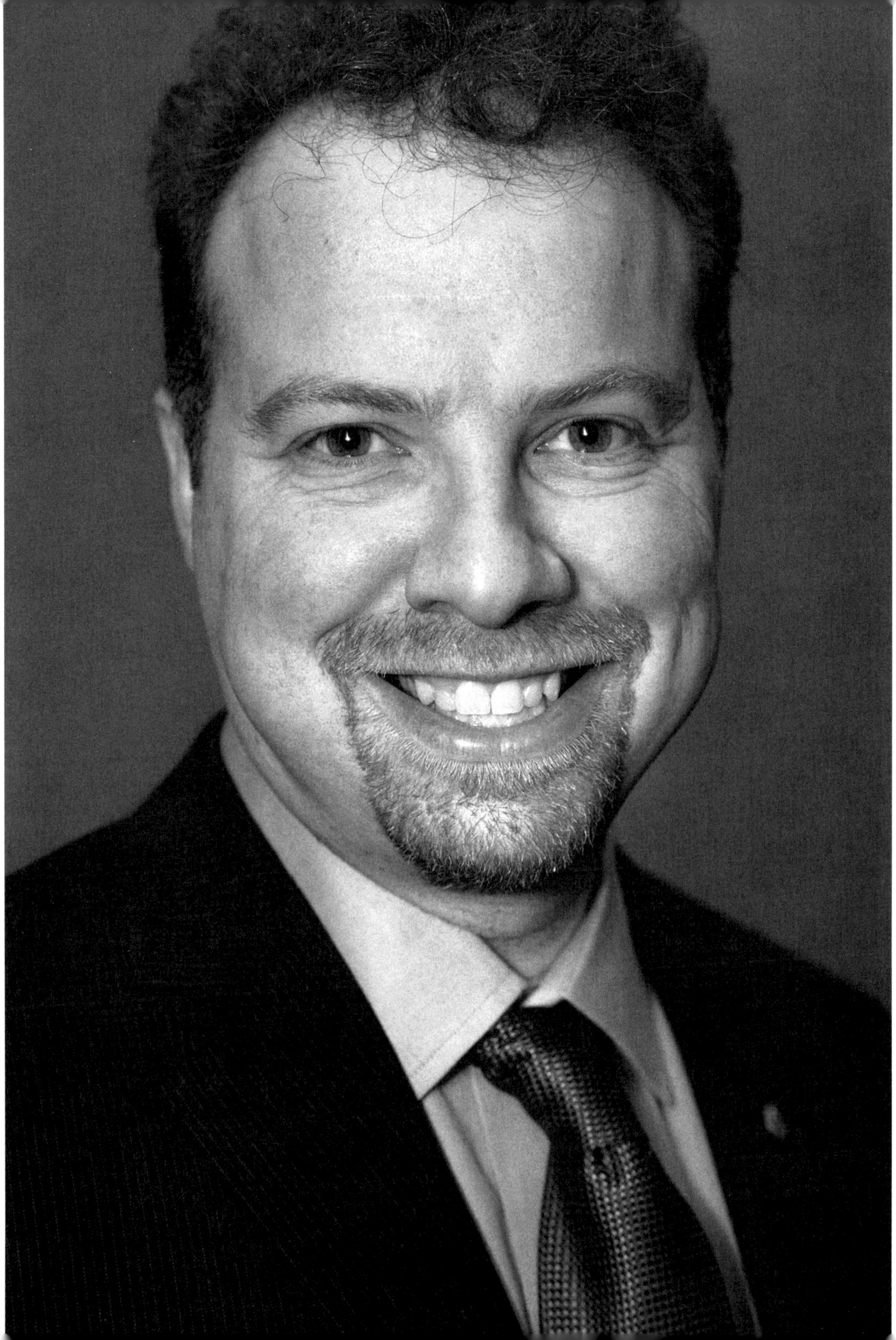

Heinrich Rohrer *1933 †2013

The Nobel Prize in Physics 1986 was divided, one half awarded to Ernst Ruska "for his fundamental work in electron optics, and for the design of the first electron microscope", the other half jointly to Gerd Binnig and Heinrich Rohrer "for their design of the scanning tunneling microscope."

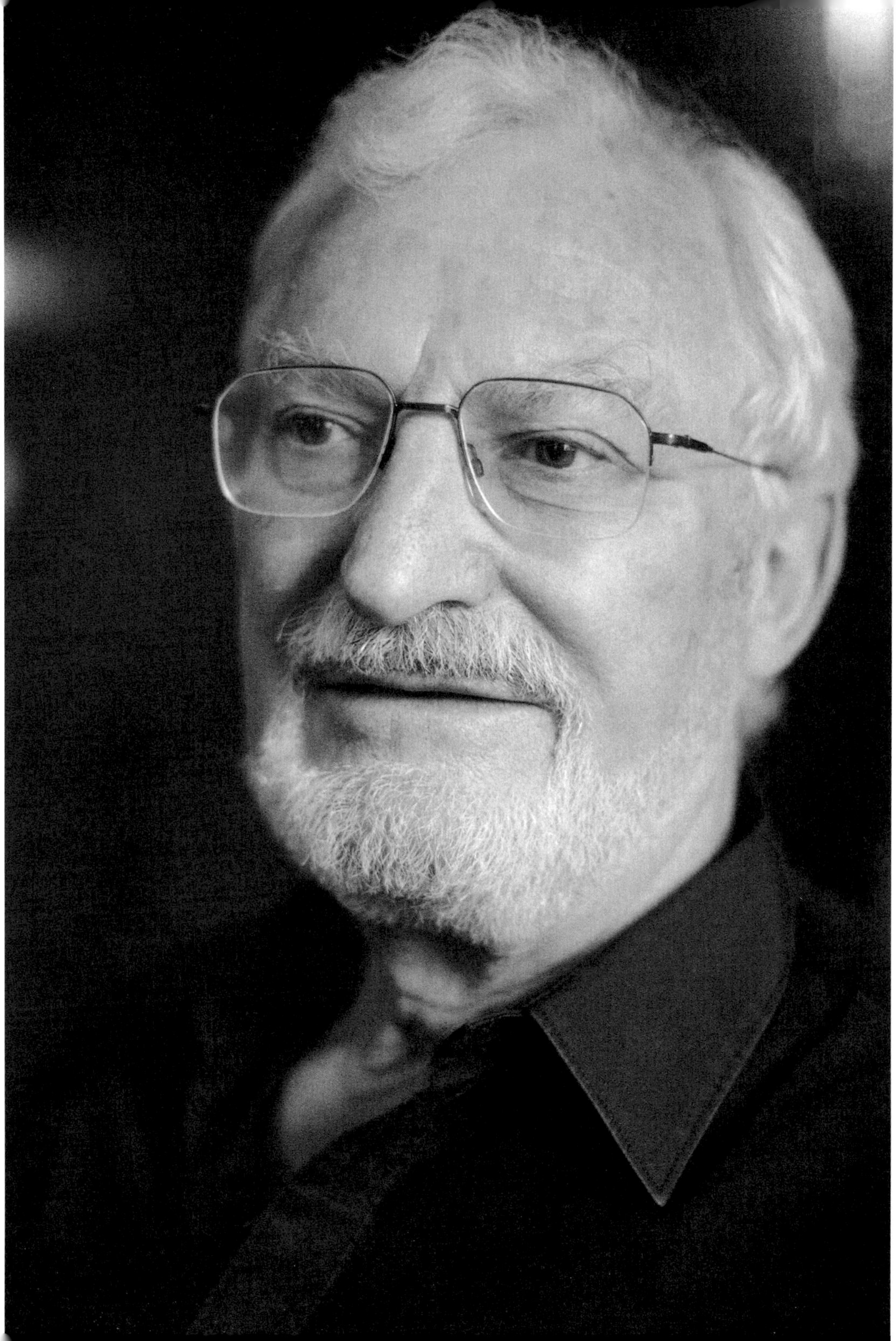

Carlo Rubbia *1934

The Nobel Prize in Physics 1984 was awarded jointly to Carlo Rubbia and Simon van der Meer "for their decisive contributions to the large project, which led to the discovery of the field particles W and Z, communicators of weak interaction."

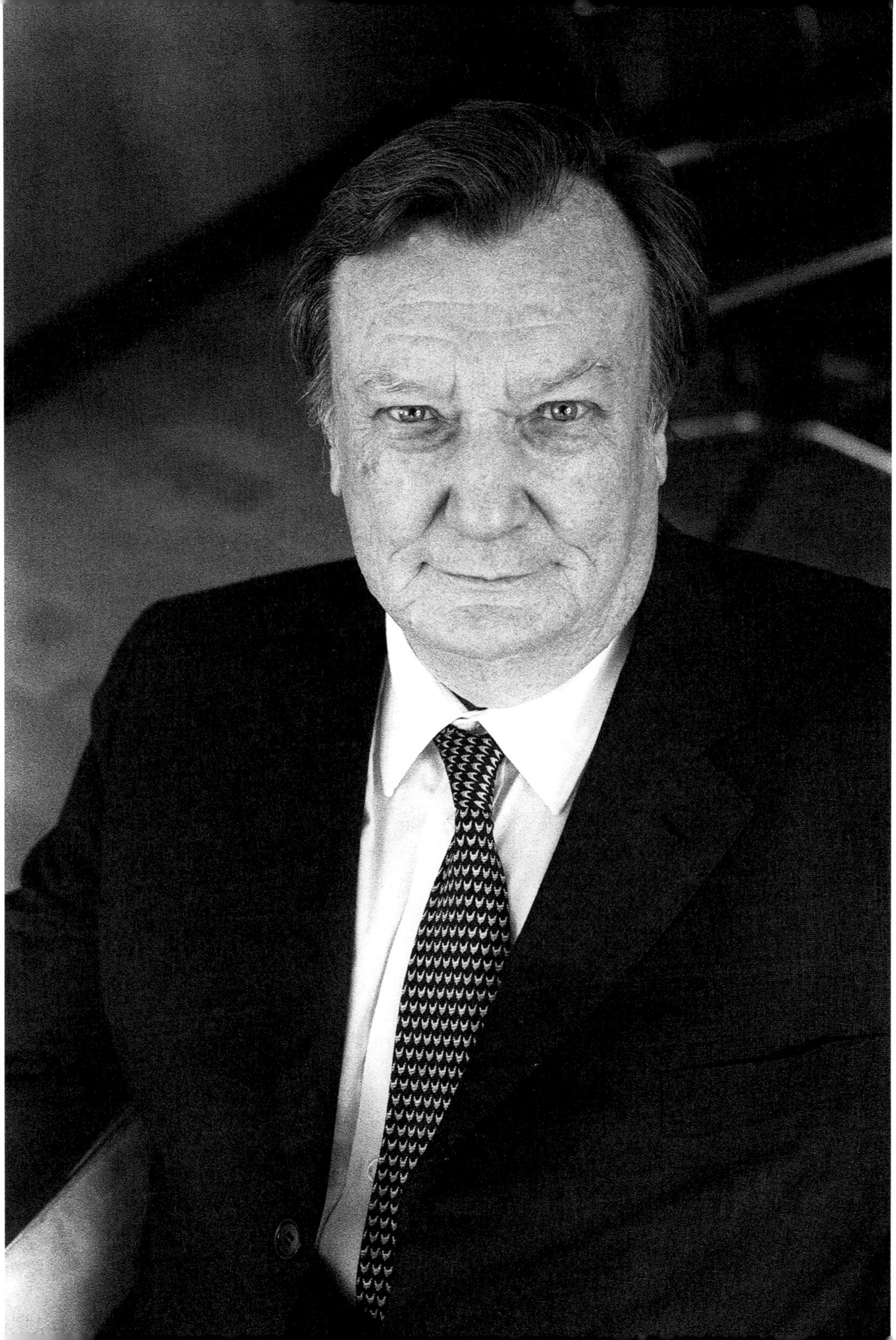

Brian P. Schmidt *1967

The Nobel Prize in Physics 2011 was divided, one half awarded to Saul Perlmutter, the other half jointly to Brian P. Schmidt and Adam G. Riess "for the discovery of the accelerating expansion of the Universe through observations of distant supernovae."

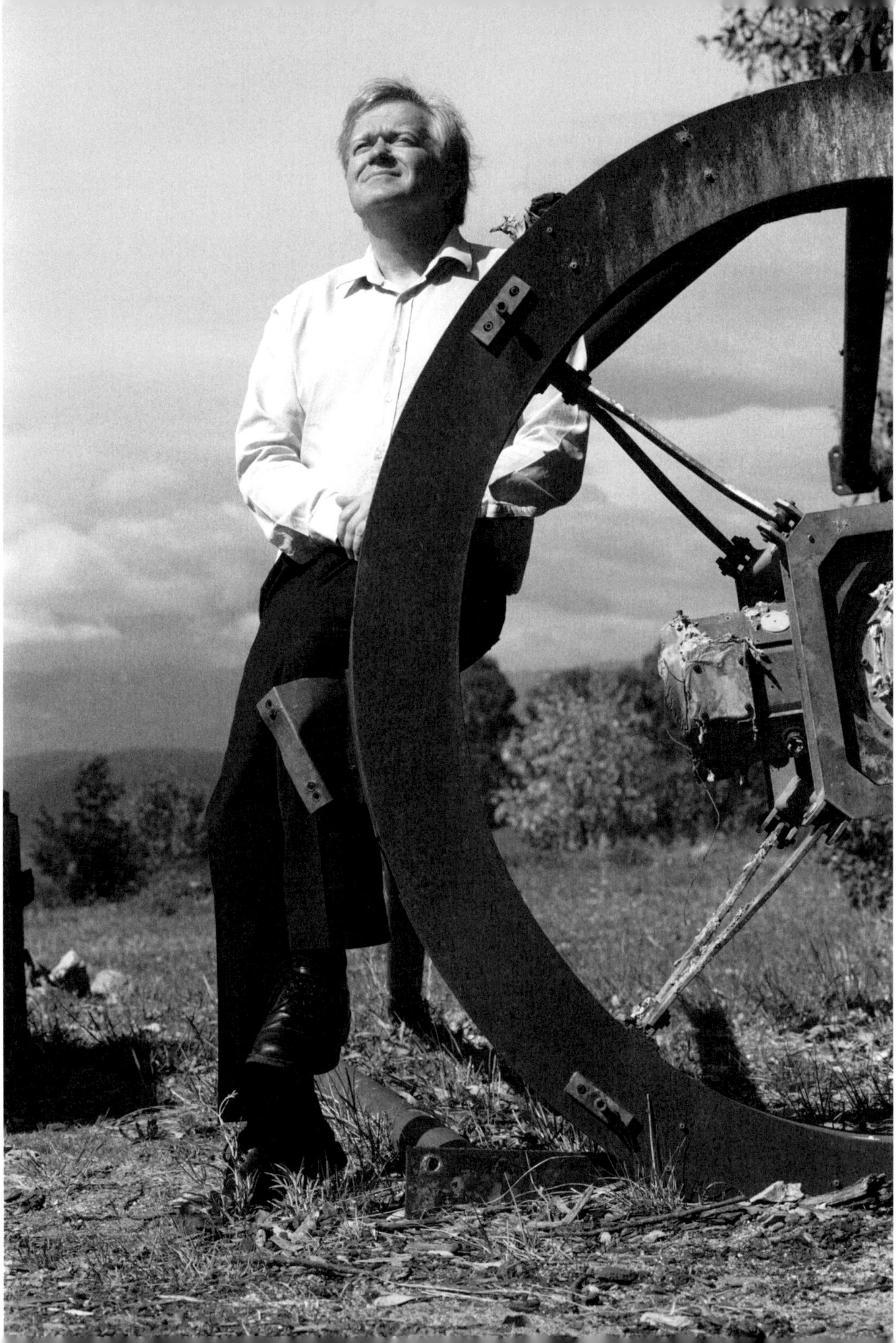

John Robert Schrieffer *1931

The Nobel Prize in Physics 1972 was awarded jointly to John Bardeen, Leon Neil Cooper and John Robert Schrieffer "for their jointly developed theory of superconductivity, usually called the BCS-theory."

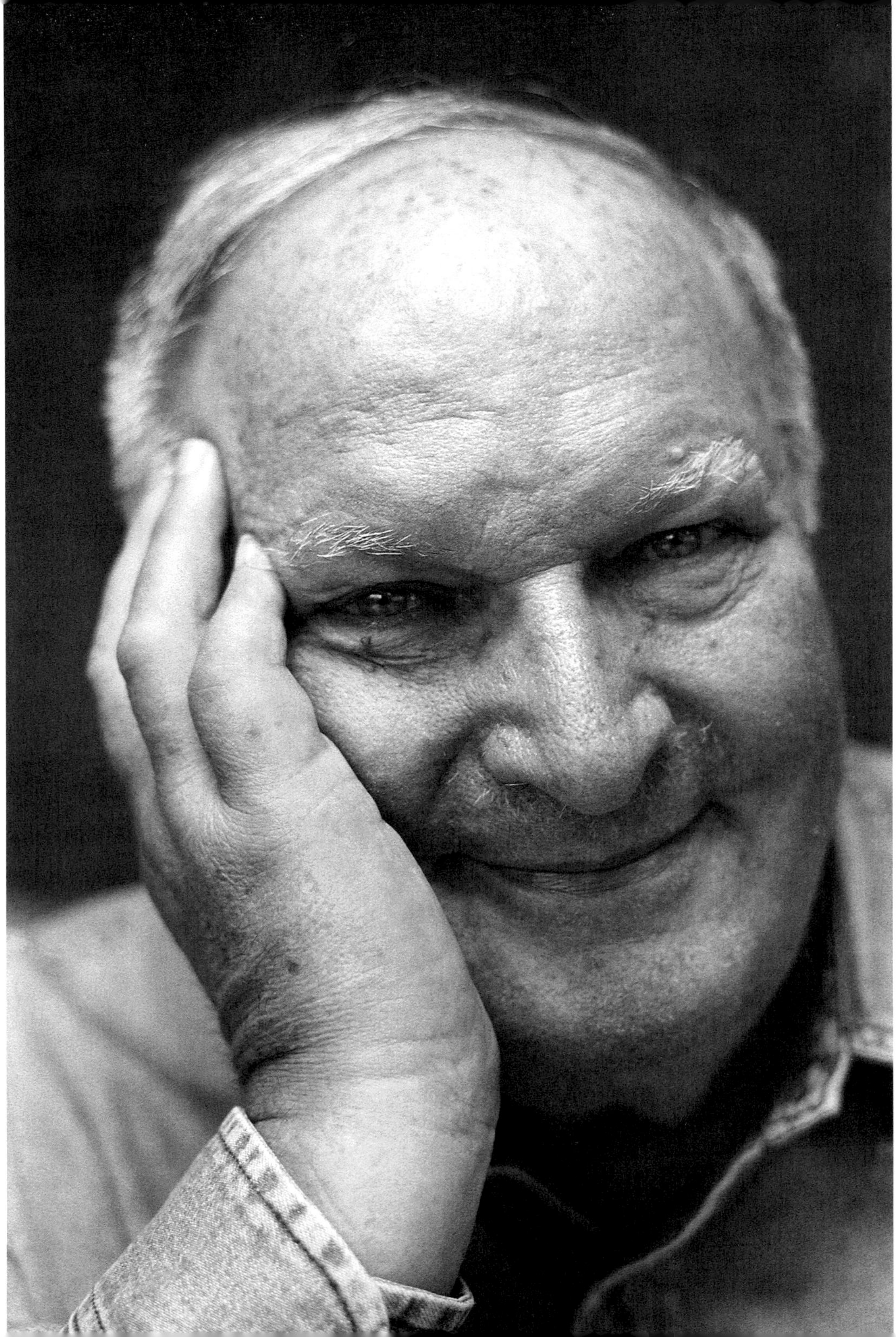

Melvin Schwartz *1932 †2006

The Nobel Prize in Physics 1988 was awarded jointly to Leon M. Lederman, Melvin Schwartz and Jack Steinberger "for the neutrino beam method and the demonstration of the doublet structure of the leptons through the discovery of the muon neutrino."

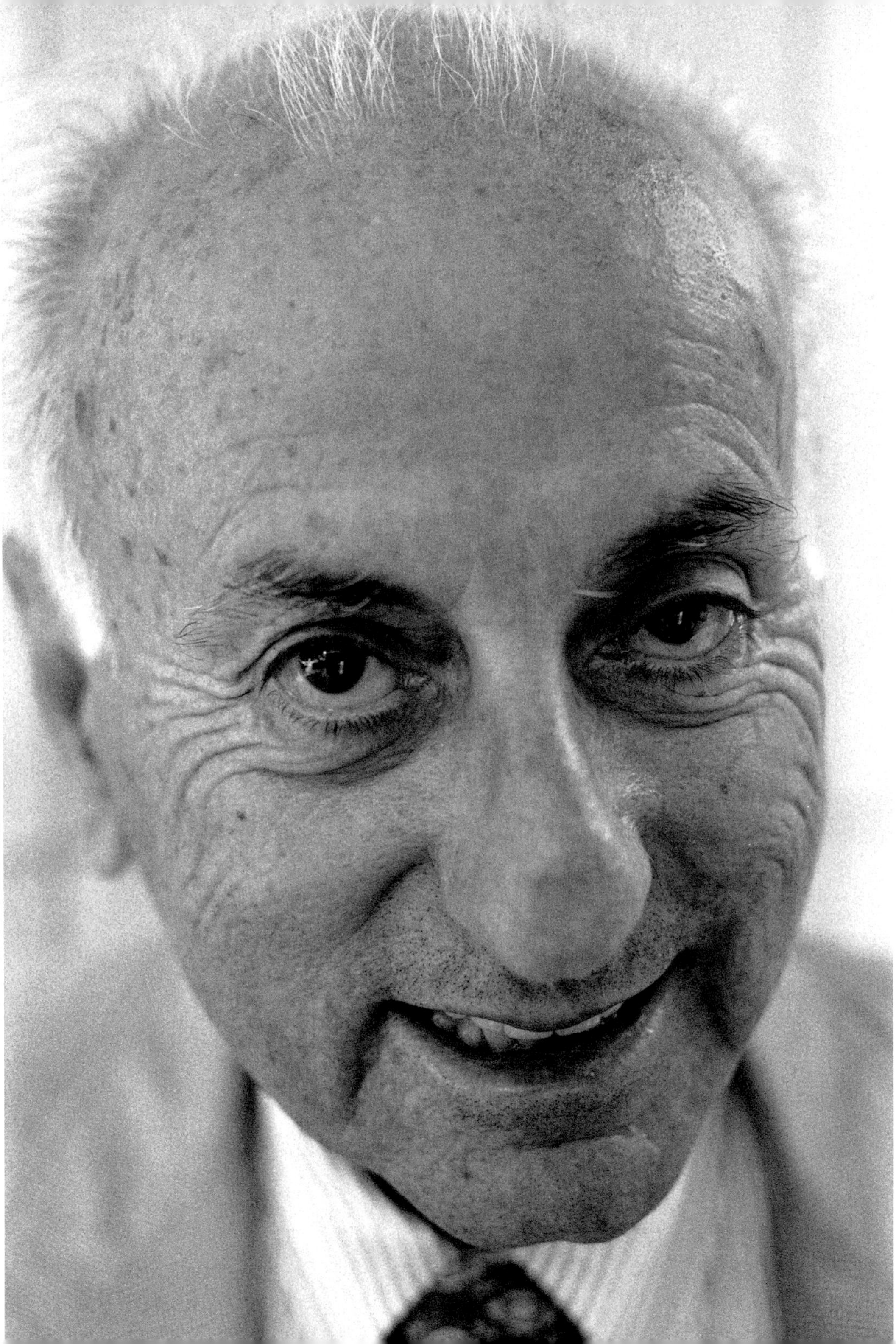

Kai M. Siegbahn *1918 †2007

The Nobel Prize in Physics 1981 was divided, one half jointly to Nicolaas Bloembergen and Arthur Leonard Schawlow "for their contribution to the development of laser spectroscopy" and the other half to Kai M. Siegbahn "for his contribution to the development of high-resolution electron spectroscopy."

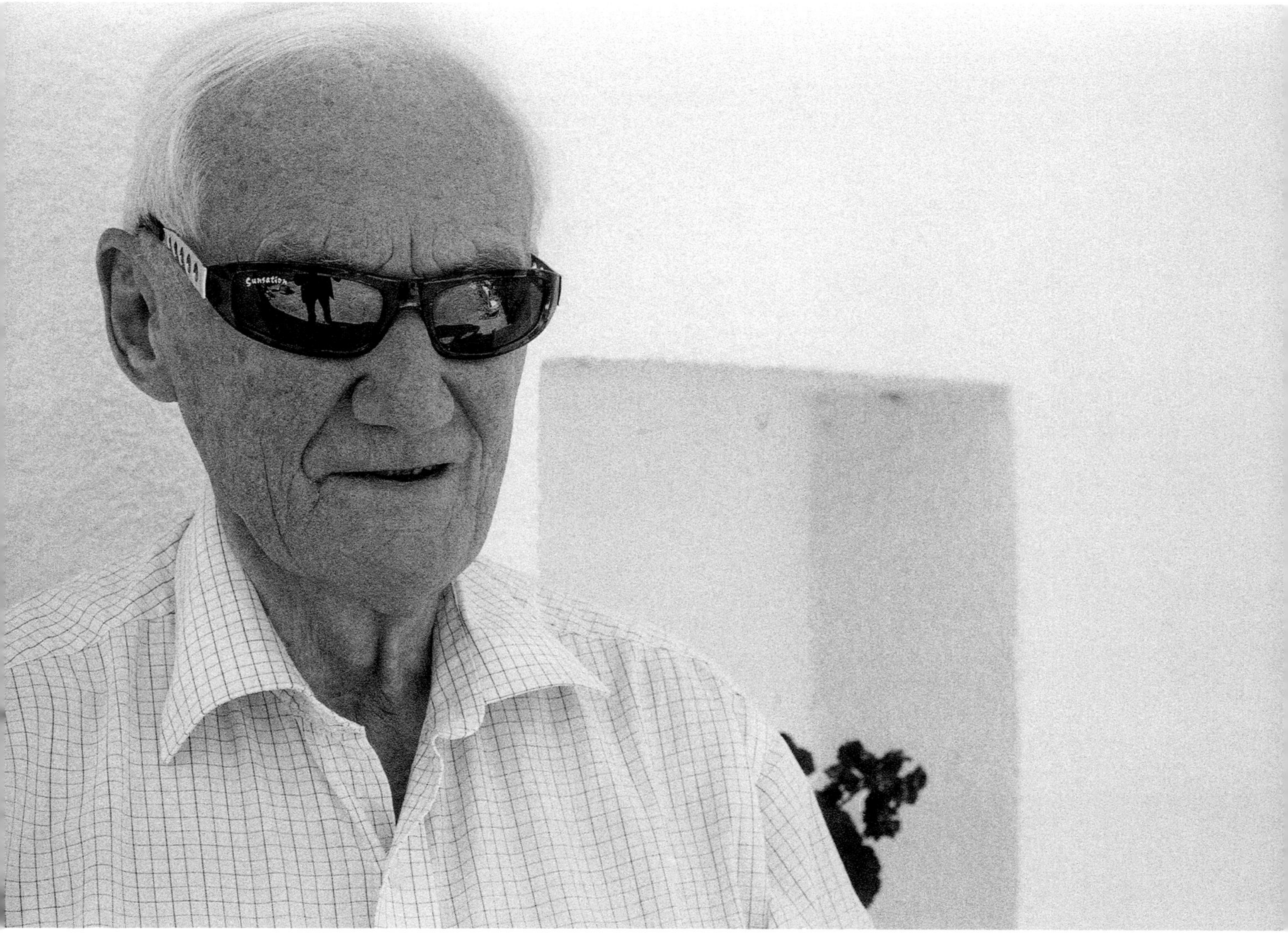
Sunsation

George E. Smith *1930

The Nobel Prize in Physics 2009 was divided, one half awarded to Charles Kuen Kao "for groundbreaking achievements concerning the transmission of light in fibers for optical communication", the other half jointly to Willard S. Boyle and George E. Smith "for the invention of an imaging semiconductor circuit – the CCD sensor."

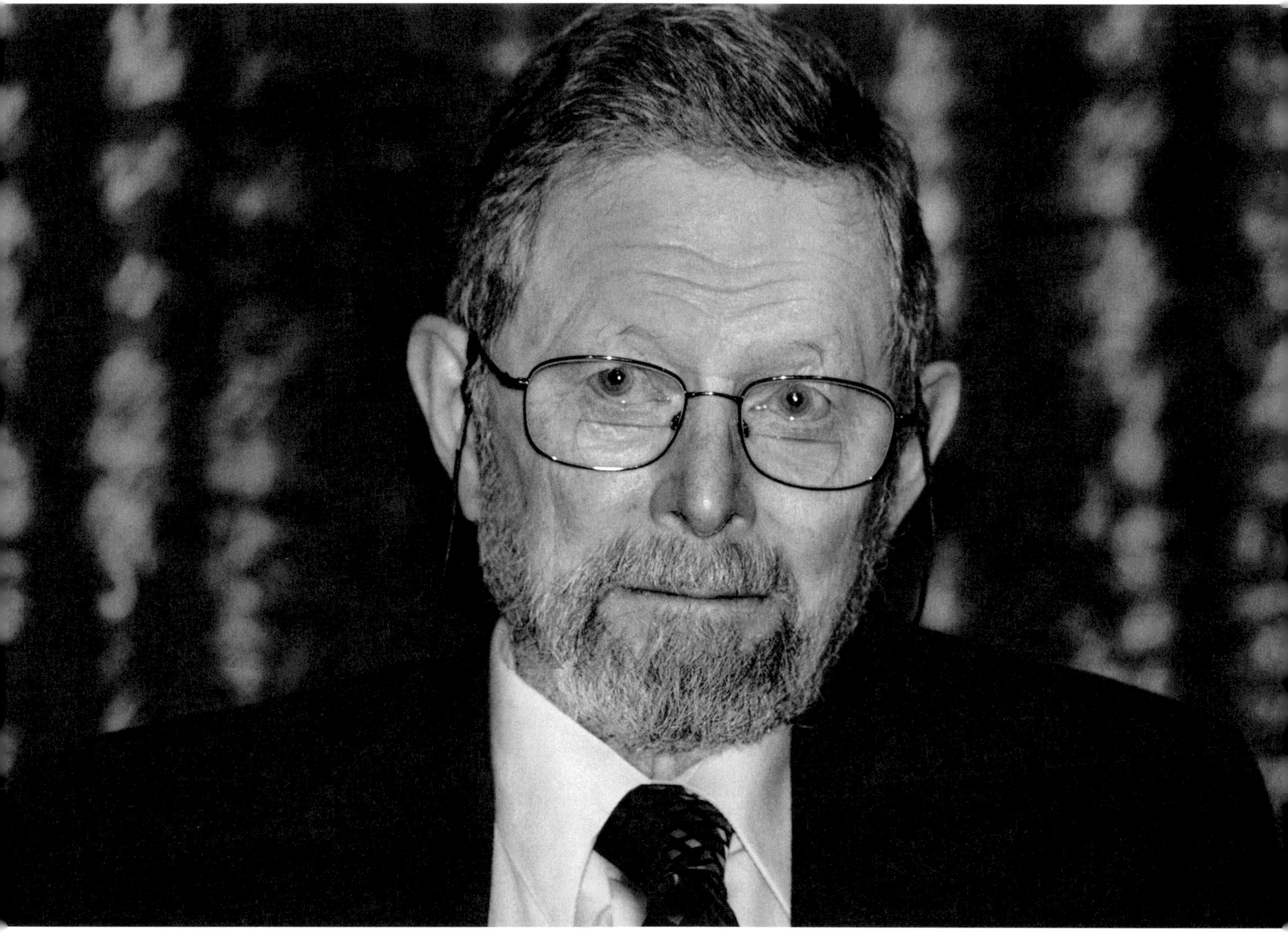

George F. Smoot *1945

The Nobel Prize in Physics 2006 was awarded jointly to John C. Mather and George F. Smoot "for their discovery of the blackbody form and anisotropy of the cosmic microwave background radiation."

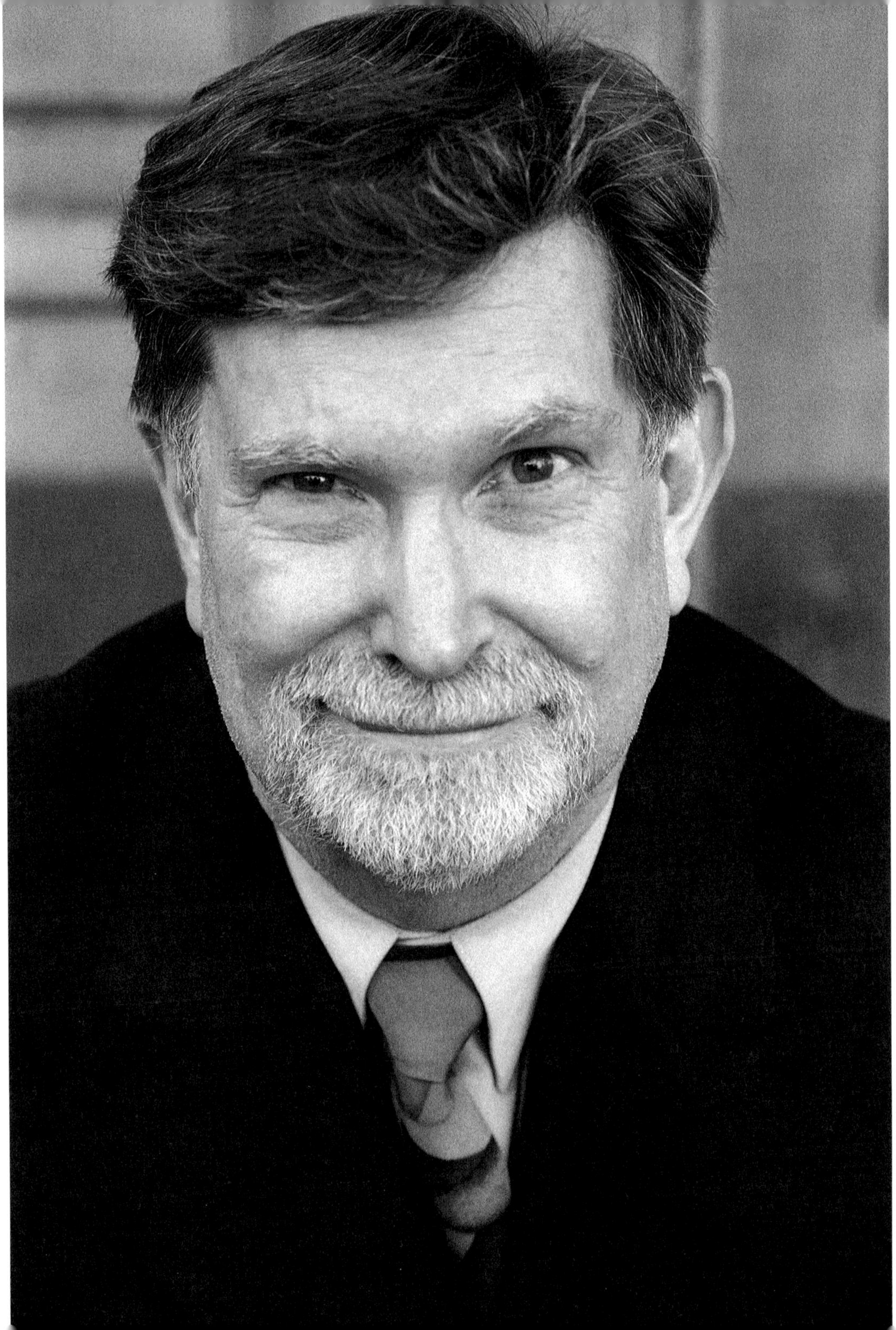

Jack Steinberger *1921

The Nobel Prize in Physics 1988 was awarded jointly to Leon M. Lederman, Melvin Schwartz and Jack Steinberger "for the neutrino beam method and the demonstration of the doublet structure of the leptons through the discovery of the muon neutrino."

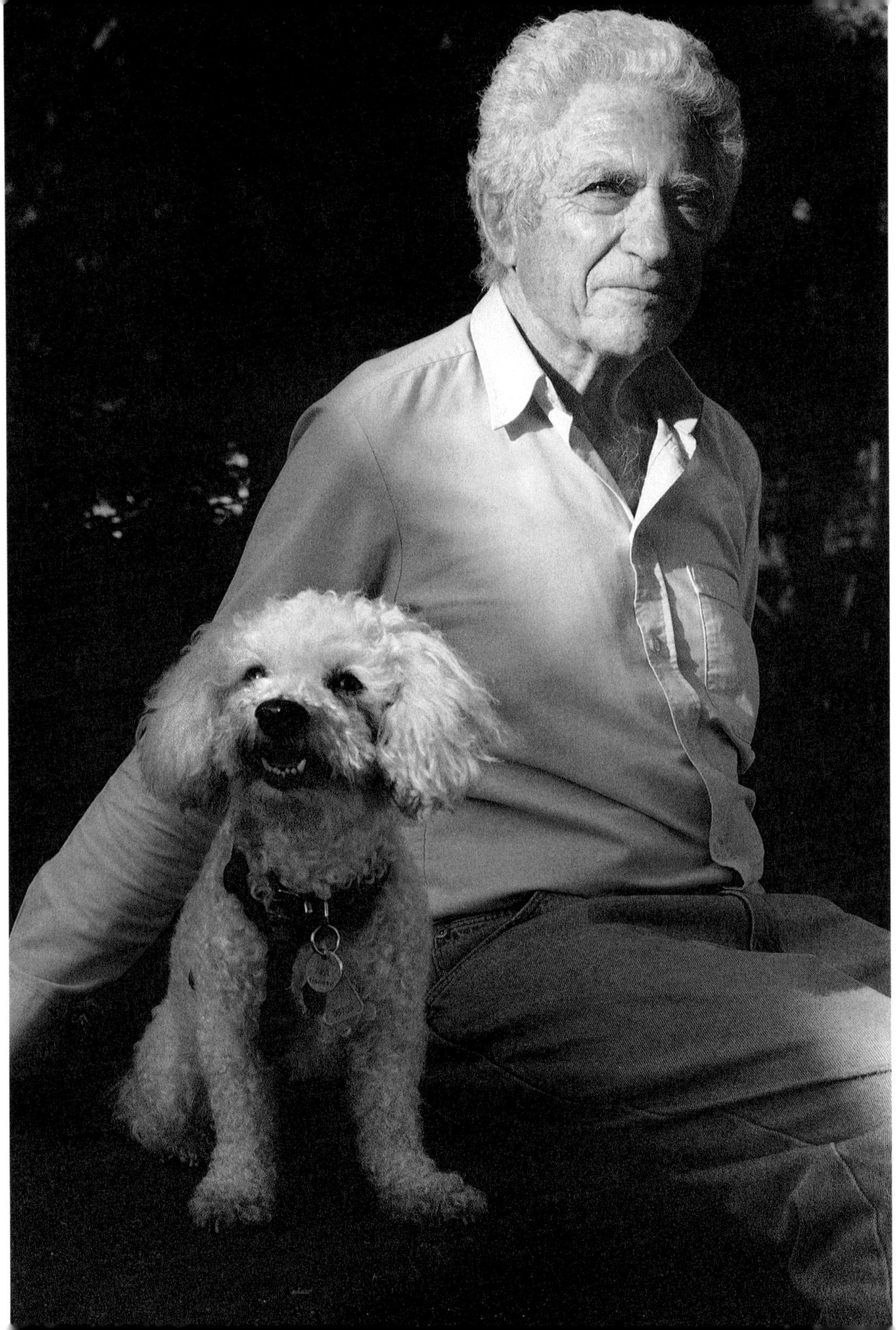

Horst L. Störmer *1949

The Nobel Prize in Physics 1998 was awarded jointly to Robert B. Laughlin, Horst L. Störmer and Daniel C. Tsui "for their discovery of a new form of quantum fluid with fractionally charged excitations."

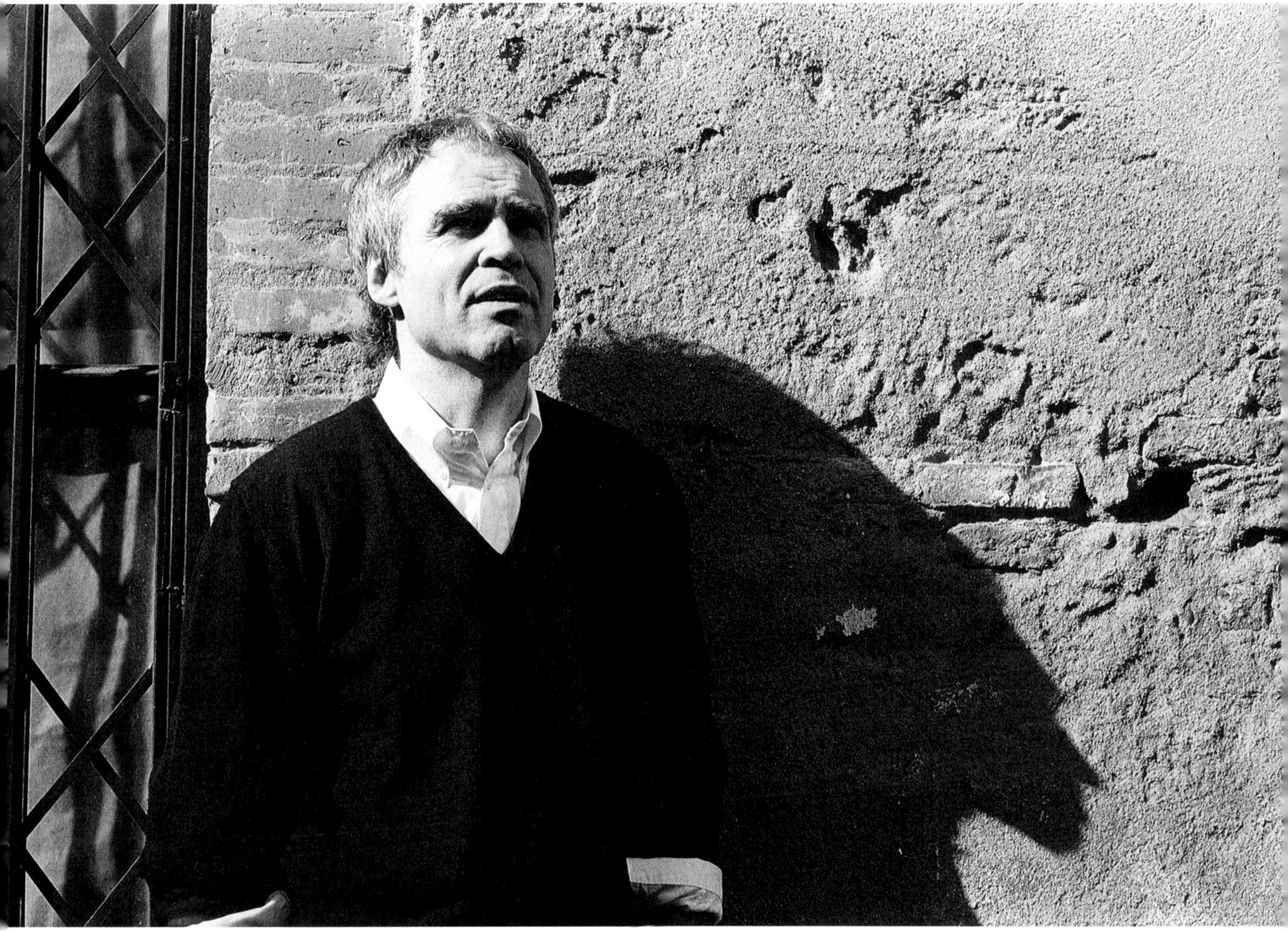

Donna Strickland *1959

The Nobel Prize in Physics 2018 was awarded "for groundbreaking inventions in the field of laser physics" with one half to Arthur Ashkin "for the optical tweezers and their application to biological systems", the other half jointly to Gérard Mourou and Donna Strickland "for their method of generating high-intensity, ultra-short optical pulses."

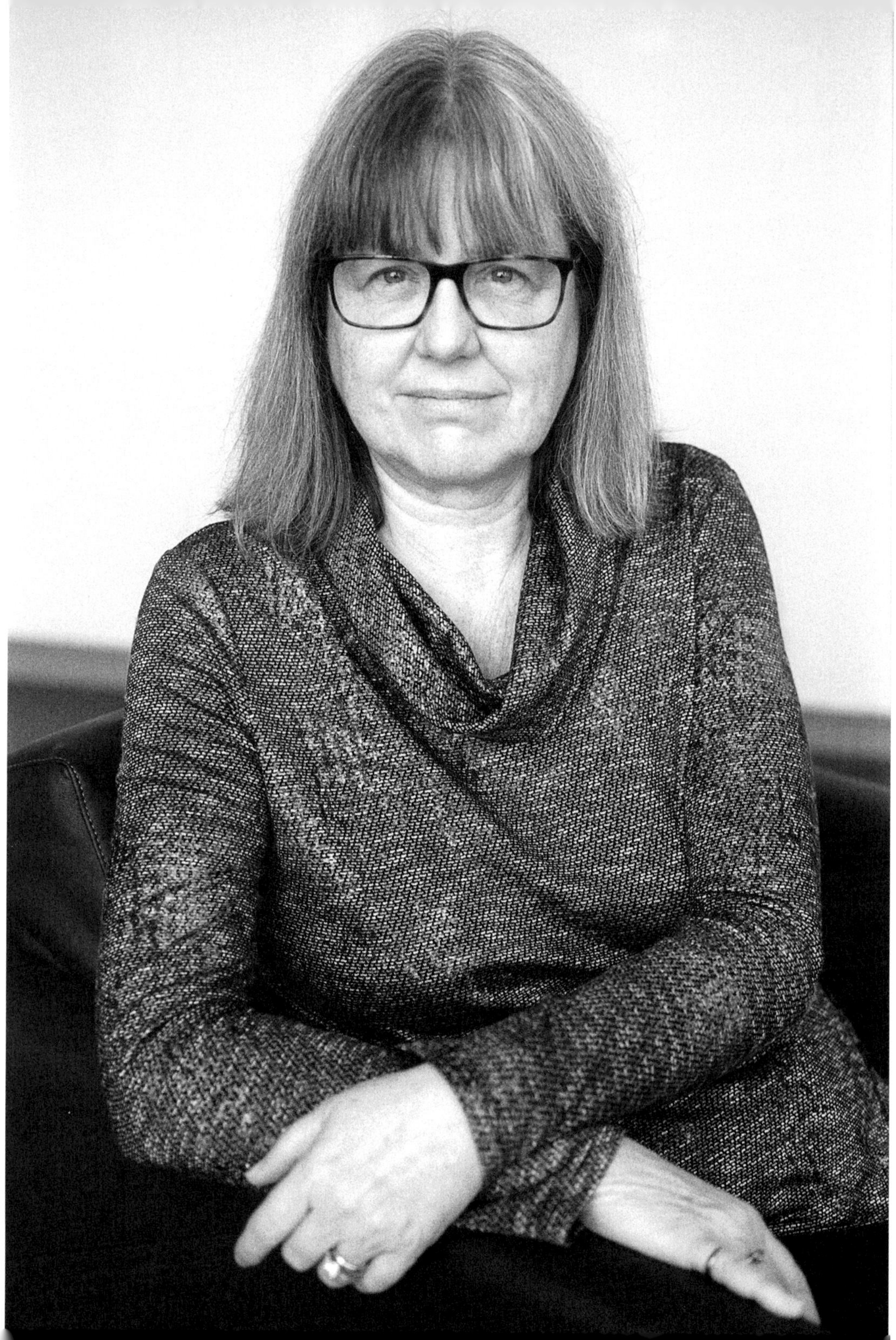

Joseph H. Taylor Jr. *1941

The Nobel Prize in Physics 1993 was awarded jointly to Russell A. Hulse and Joseph H. Taylor Jr. "for the discovery of a new type of pulsar, a discovery that has opened up new possibilities for the study of gravitation."

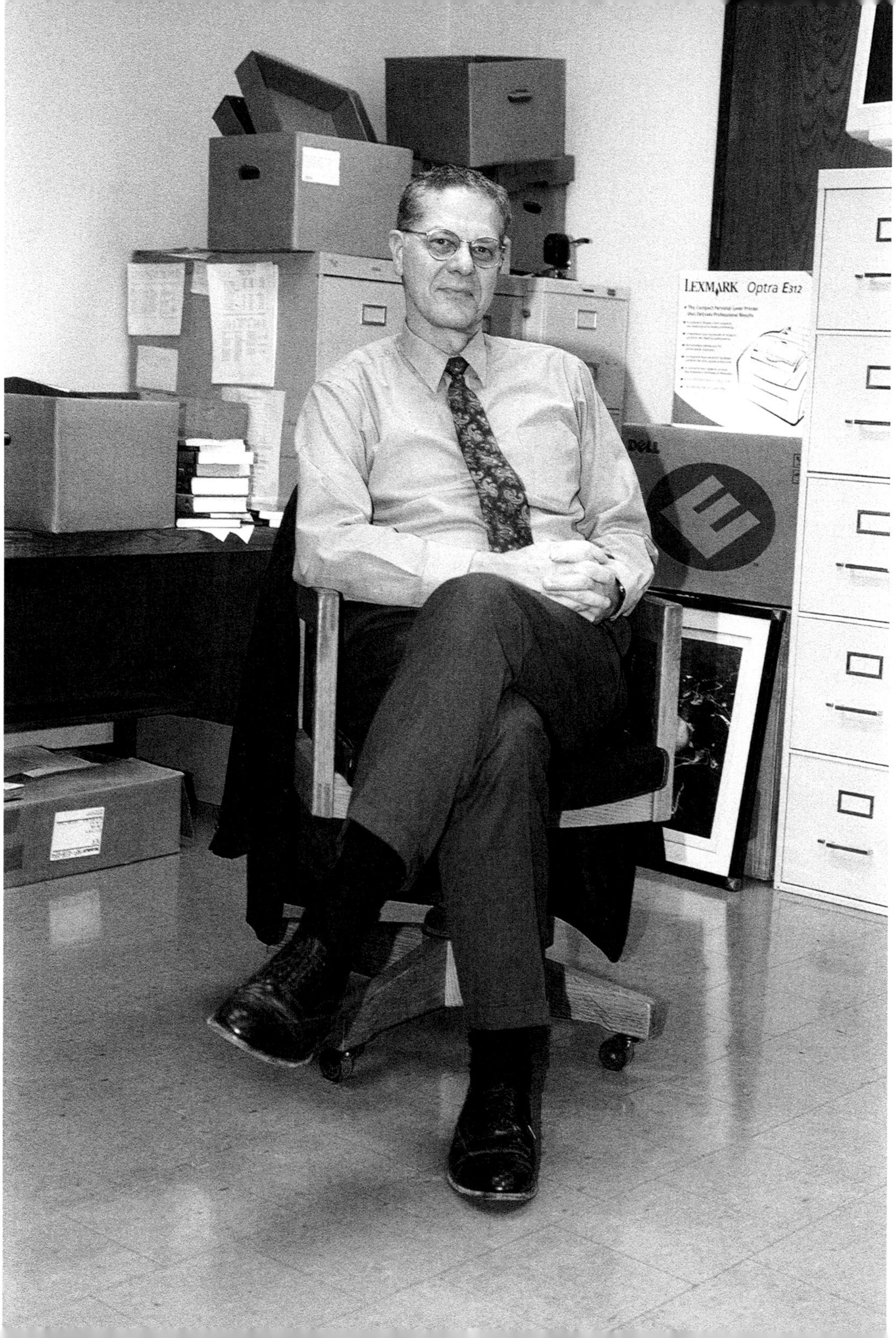
LEXMARK Optra E312

Richard E. Taylor *1929 †2018

The Nobel Prize in Physics 1990 was awarded jointly to Jerome I. Friedman, Henry W. Kendall and Richard E. Taylor "for their pioneering investigations concerning deep inelastic scattering of electrons on protons and bound neutrons, which have been of essential importance for the development of the quark model in particle physics."

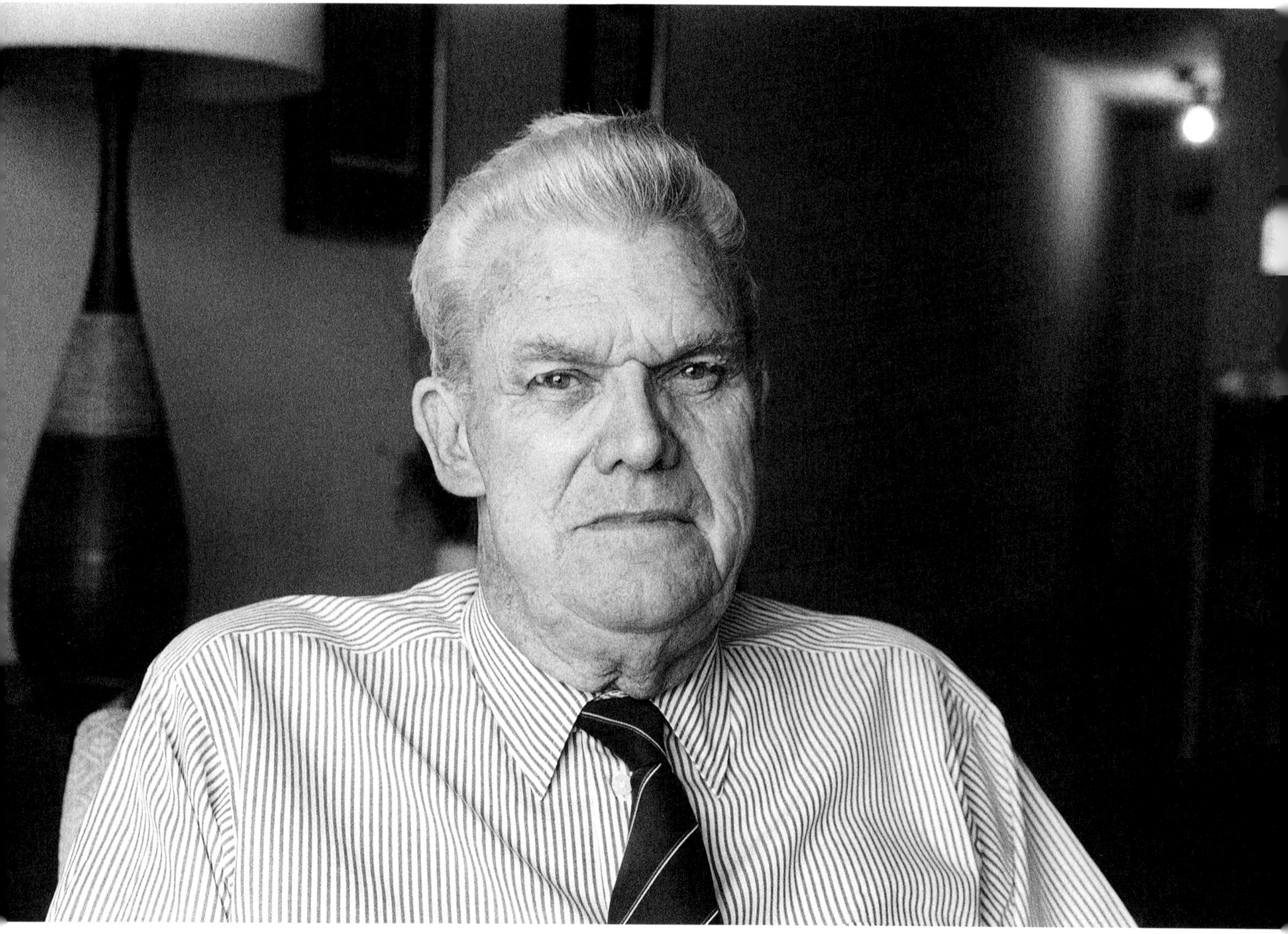

Kip S. Thorne *1940

The Nobel Prize in Physics 2017 was divided, one half awarded to Rainer Weiss, the other half jointly to Barry C. Barish and Kip S. Thorne "for decisive contributions to the LIGO detector and the observation of gravitational waves."

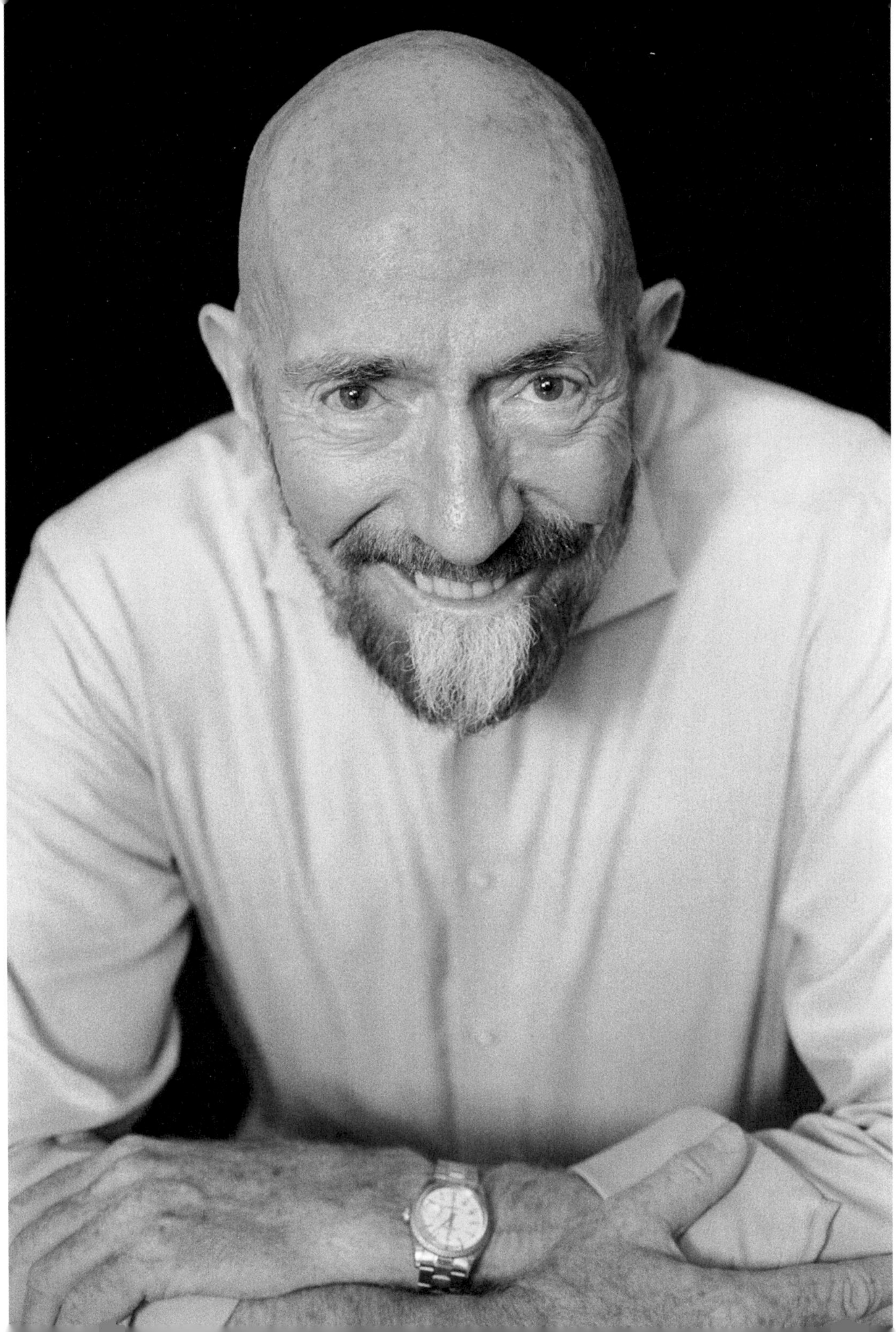

David J. Thouless *1934 †2019

The Nobel Prize in Physics 2016 was divided, one half awarded to David J. Thouless, the other half jointly to F. Duncan M. Haldane and J. Michael Kosterlitz "for theoretical discoveries of topological phase transitions and topological phases of matter."

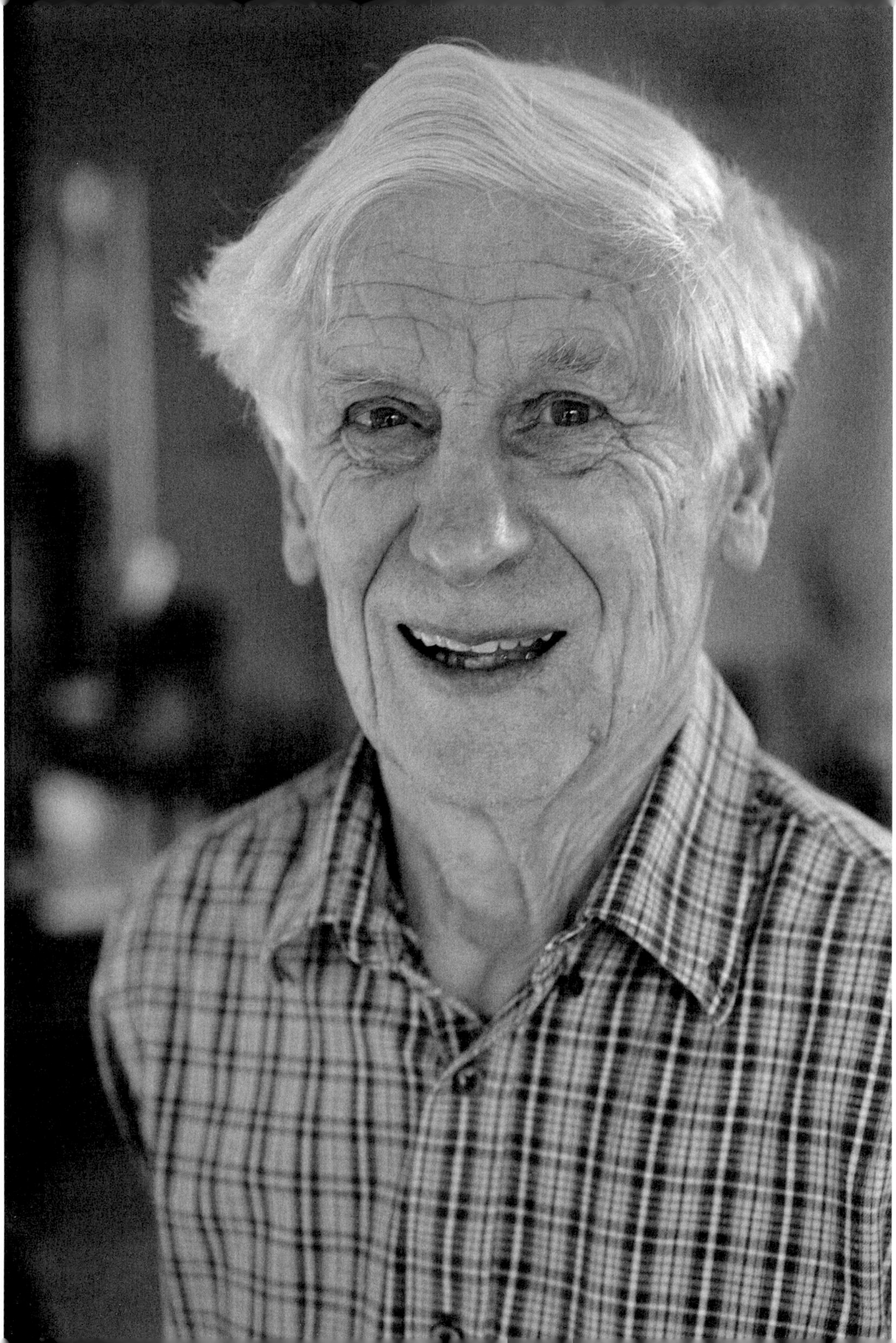

Samuel Chao Chung Ting *1936

The Nobel Prize in Physics 1976 was awarded jointly to Burton Richter and Samuel Chao Chung Ting "for their pioneering work in the discovery of a heavy elementary particle of a new kind."

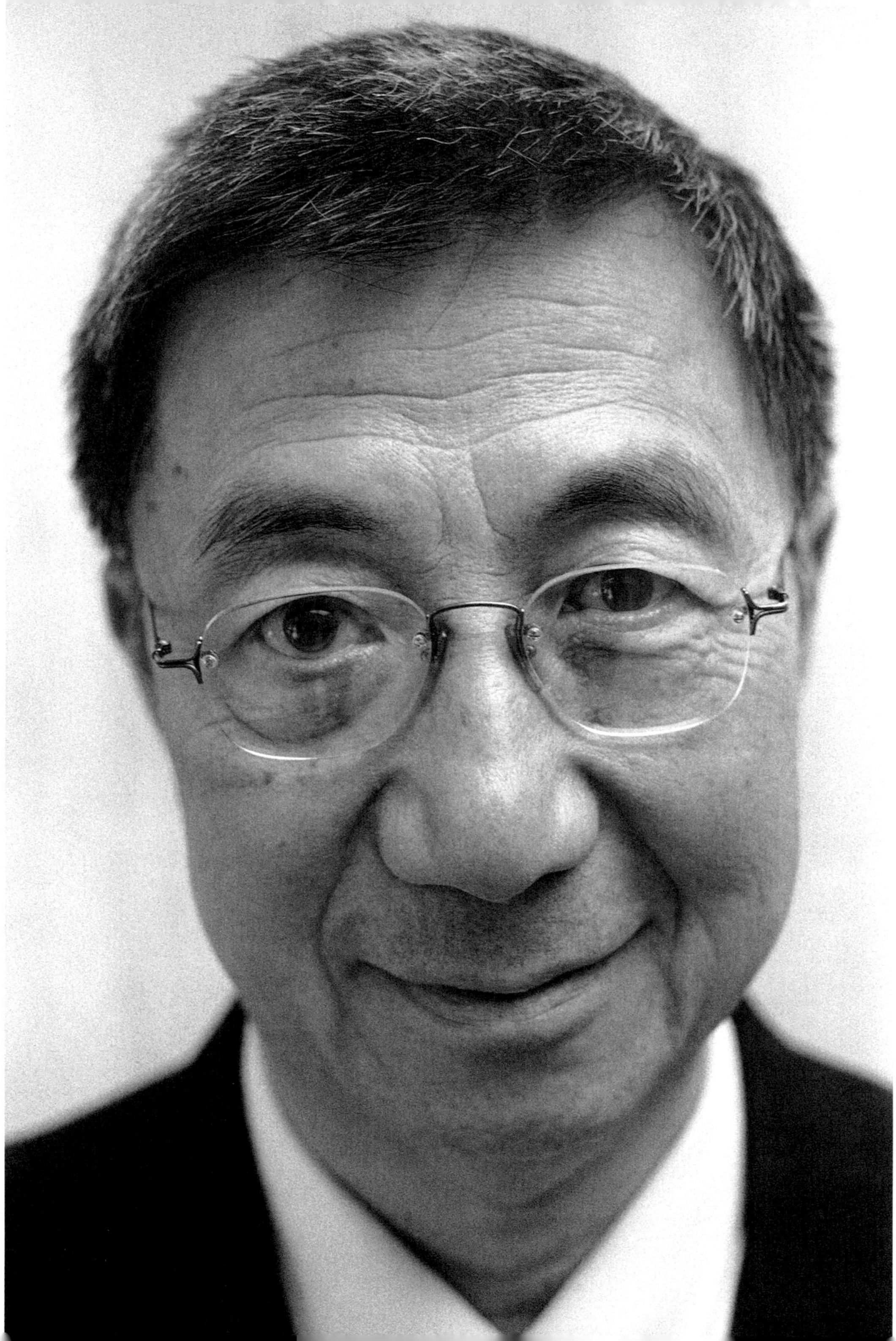

Charles Hard Townes *1915 †2015

The Nobel Prize in Physics 1964 was divided, one half awarded to Charles Hard Townes, the other half jointly to Nicolay Gennadiyevich Basov and Aleksandr Mikhailovich Prokhorov "for fundamental work in the field of quantum electronics, which has led to the construction of oscillators and amplifiers based on the maser-laser principle."

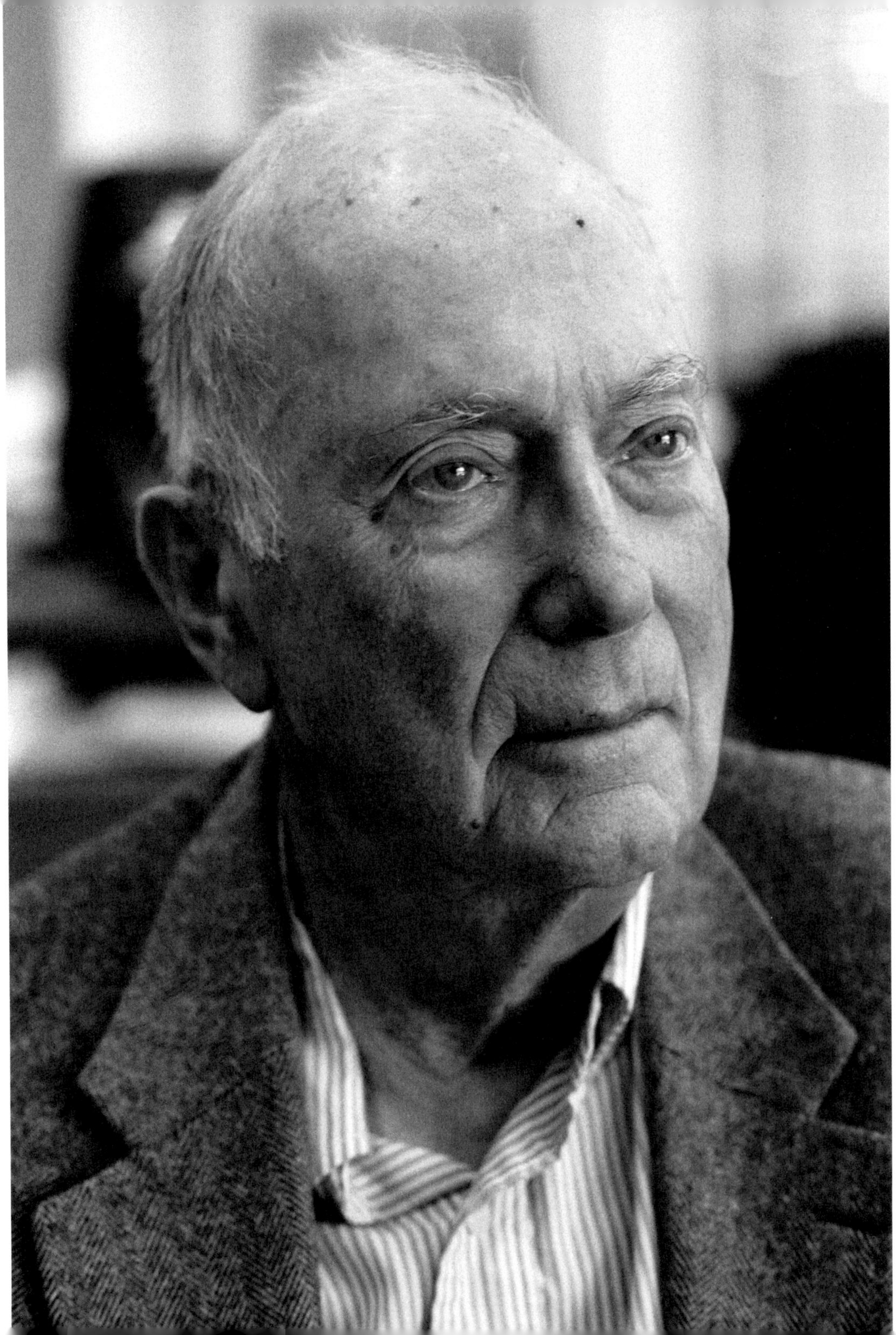

Daniel C. Tsui *1939

The Nobel Prize in Physics 1998 was awarded jointly to Robert B. Laughlin, Horst L. Störmer and Daniel C. Tsui "for their discovery of a new form of quantum fluid with fractionally charged excitations."

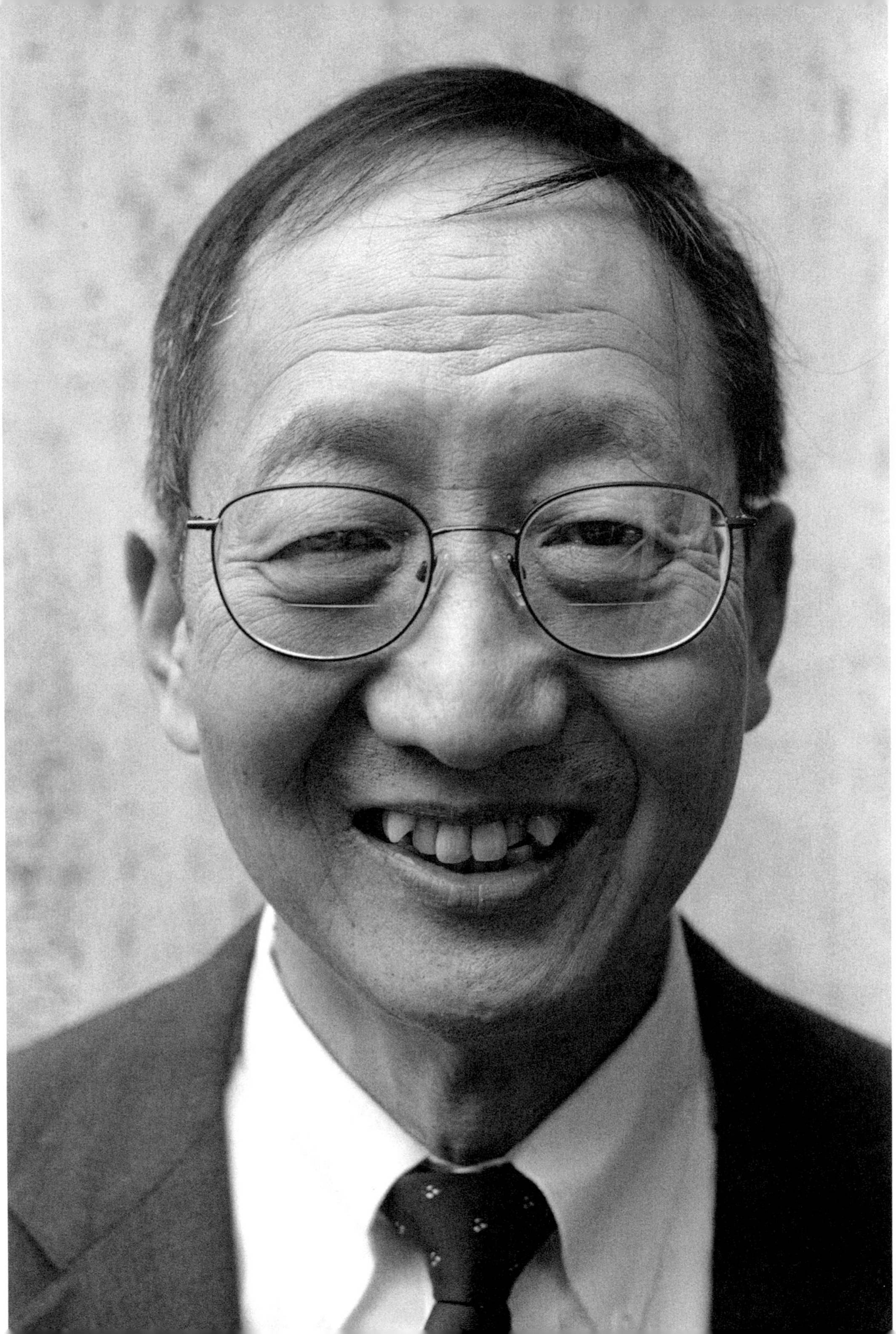

Martinus J.G. Veltman *1931

The Nobel Prize in Physics 1999 was awarded jointly to Gerardus 't Hooft and Martinus J.G. Veltman "for elucidating the quantum structure of electroweak interactions in physics."

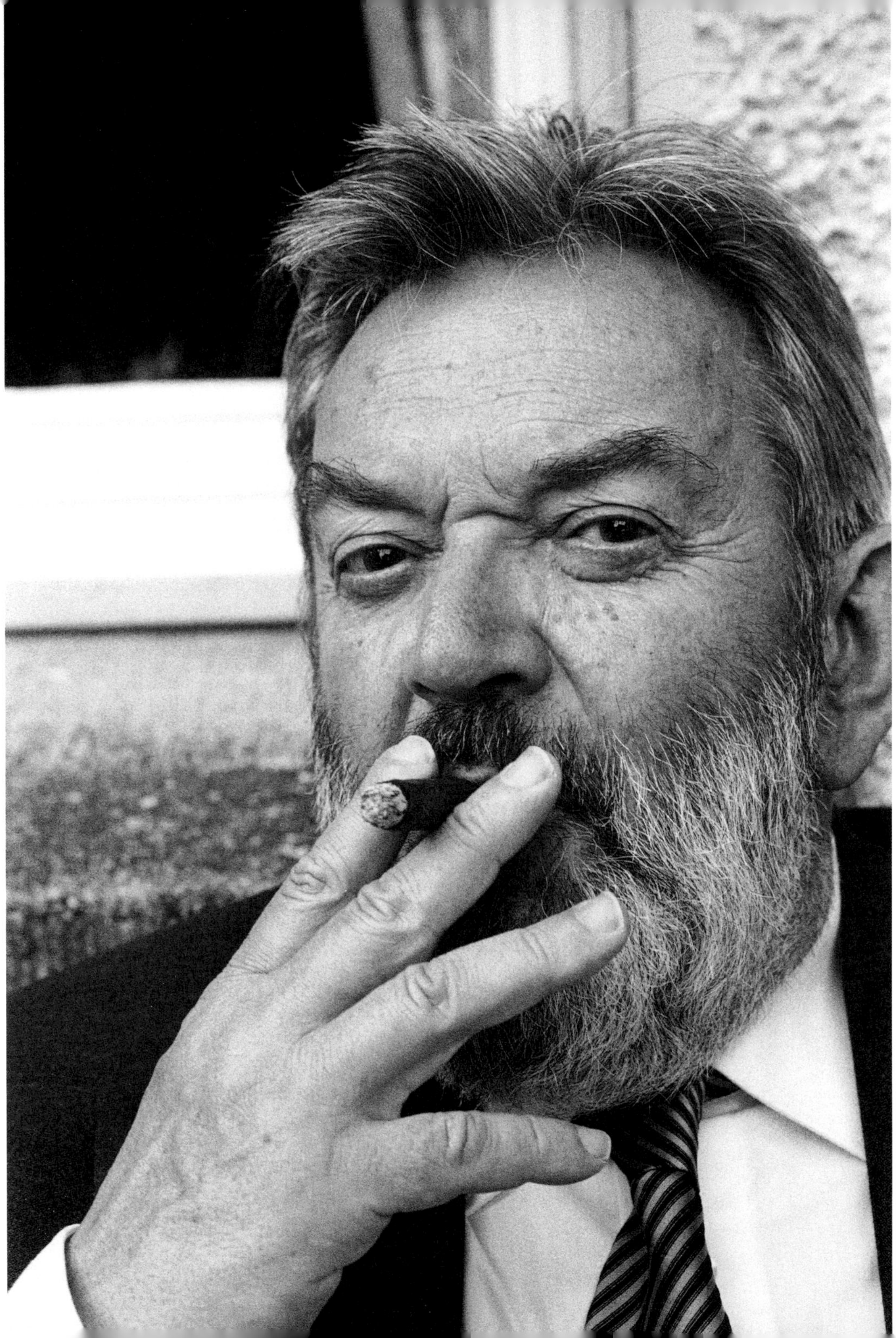

Steven Weinberg *1933

The Nobel Prize in Physics 1979 was awarded jointly to Sheldon Lee Glashow, Abdus Salam and Steven Weinberg "for their contributions to the theory of the unified weak and electromagnetic interaction between elementary particles, including, inter alia, the prediction of the weak neutral current."

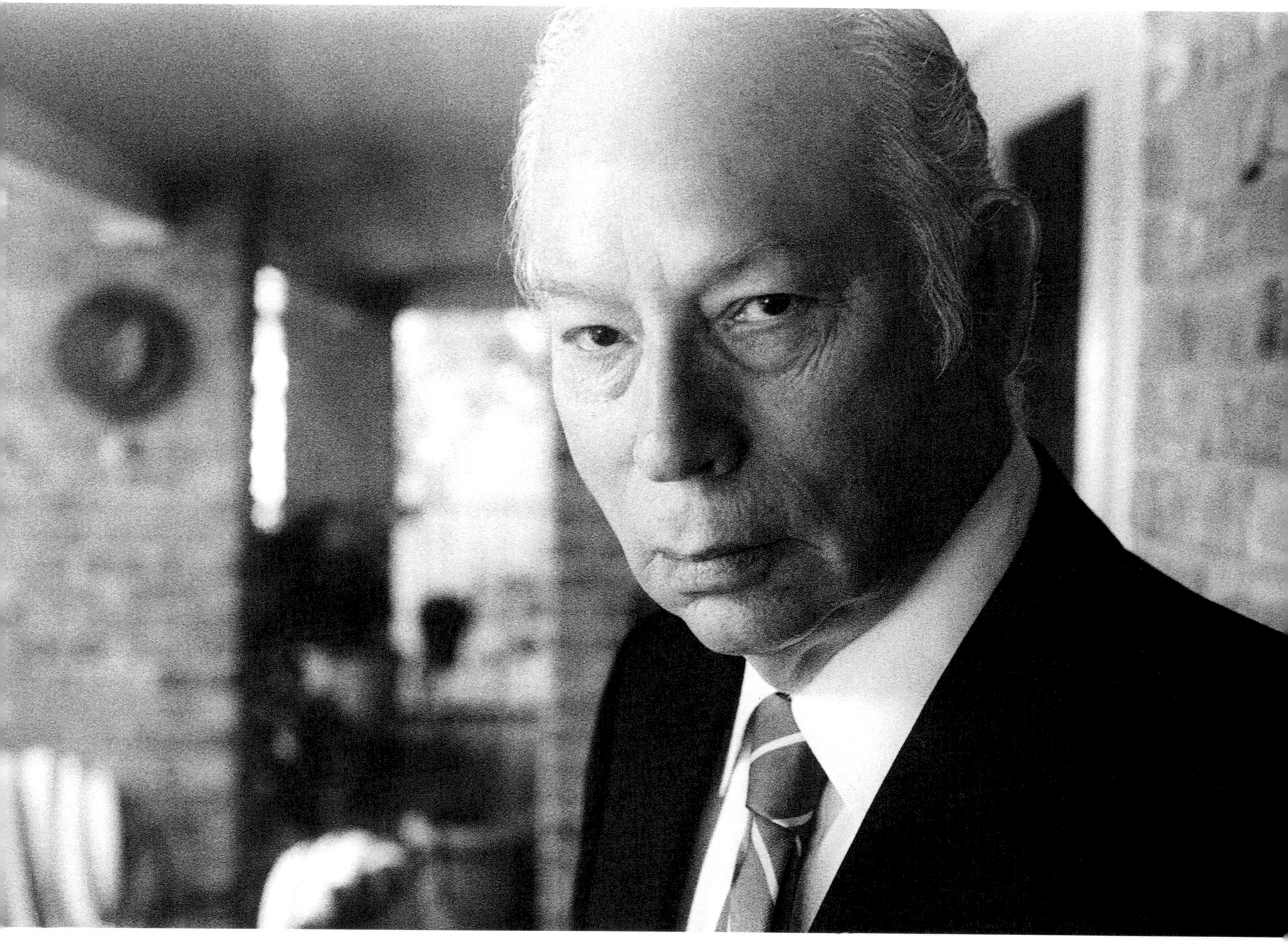

Rainer Weiss *1932

The Nobel Prize in Physics 2017 was divided, one half awarded to Rainer Weiss, the other half jointly to Barry C. Barish and Kip S. Thorne "for decisive contributions to the LIGO detector and the observation of gravitational waves."

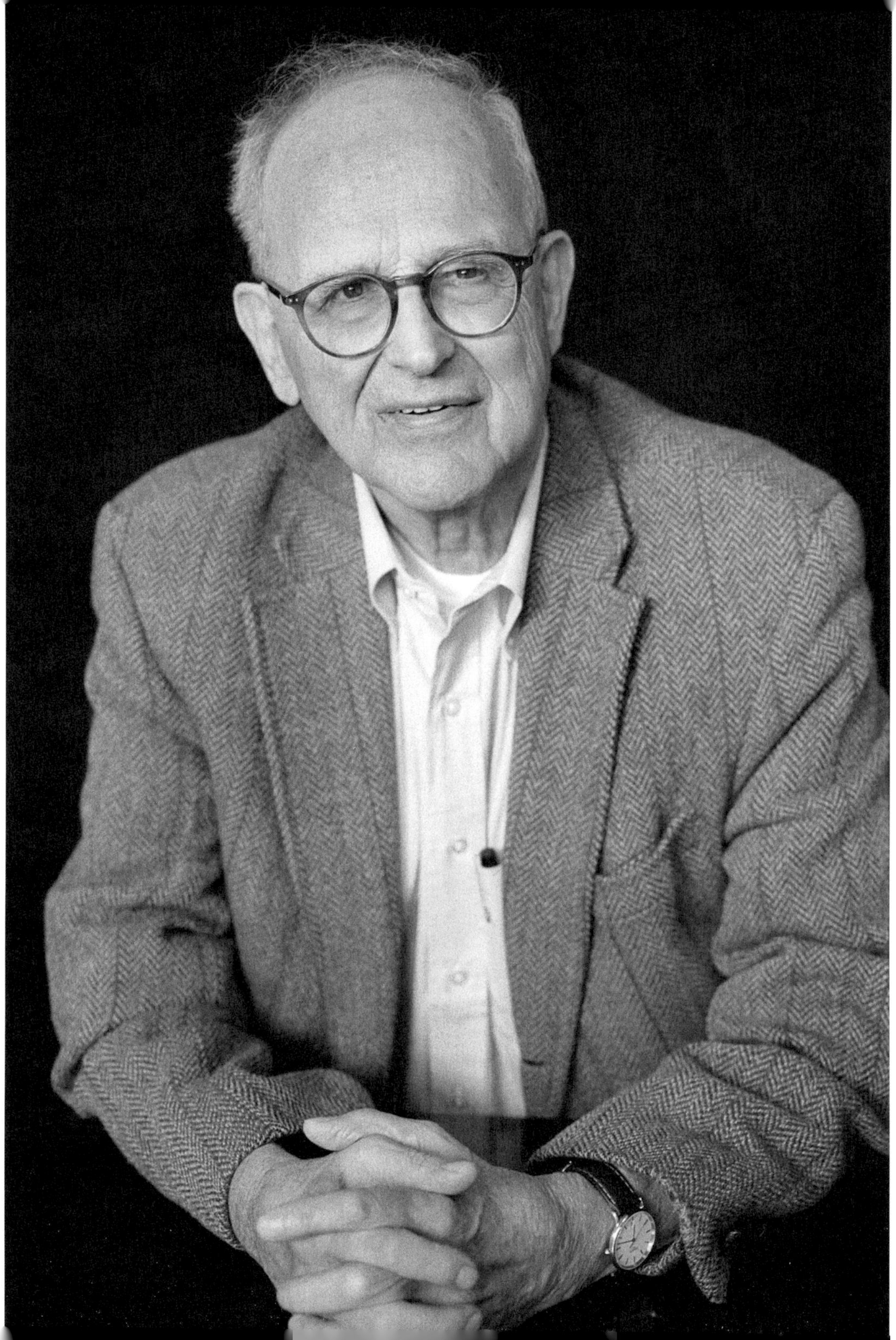

Carl E. Wieman *1951

The Nobel Prize in Physics 2001 was awarded jointly to Eric A. Cornell, Wolfgang Ketterle and Carl E. Wieman "for the achievement of Bose-Einstein condensation in dilute gases of alkali atoms, and for early fundamental studies of the properties of the condensates."

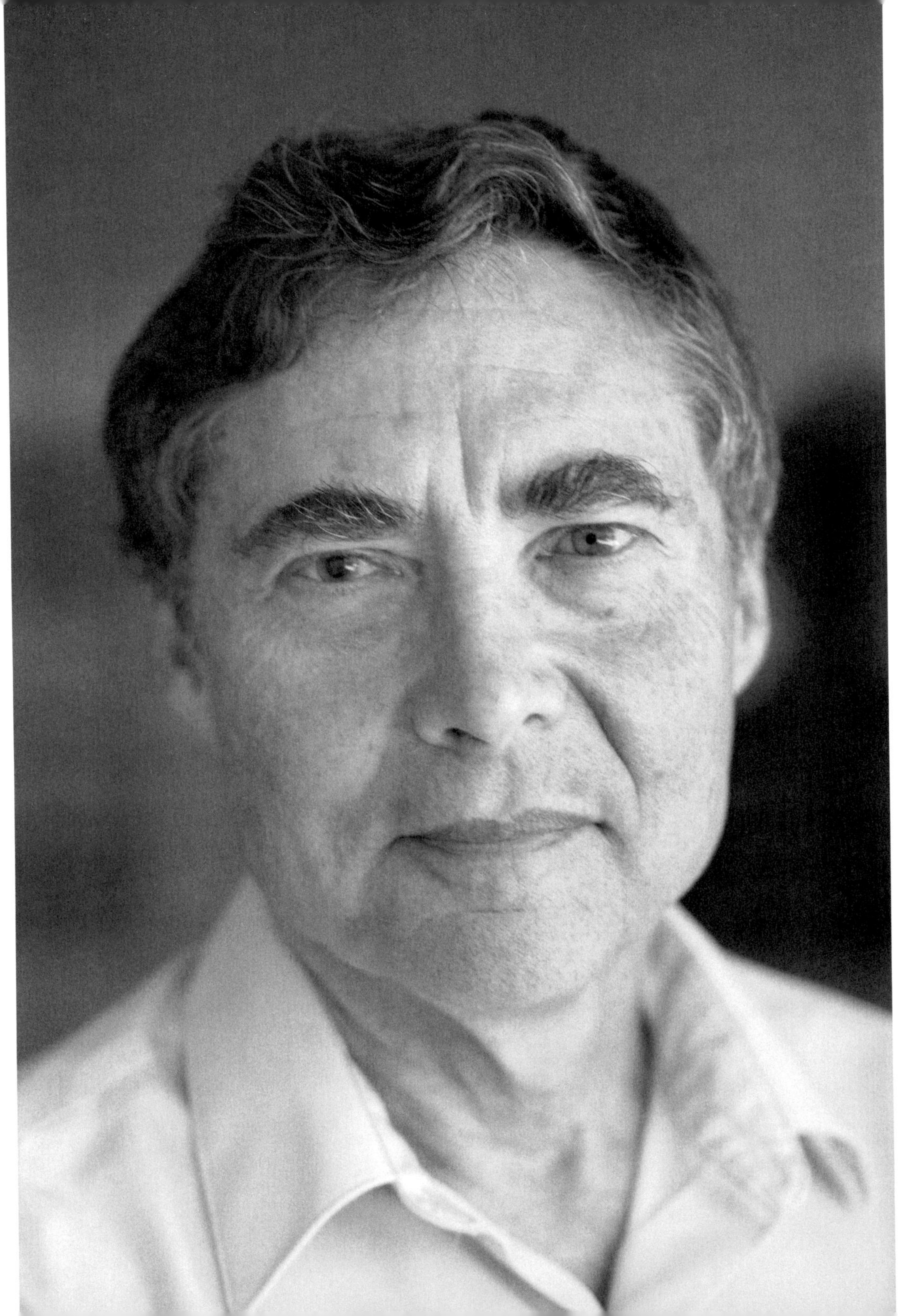

Frank Wilczek *1951

The Nobel Prize in Physics 2004 was awarded jointly to David J. Gross, H. David Politzer and Frank Wilczek "for the discovery of asymptotic freedom in the theory of the strong interaction."

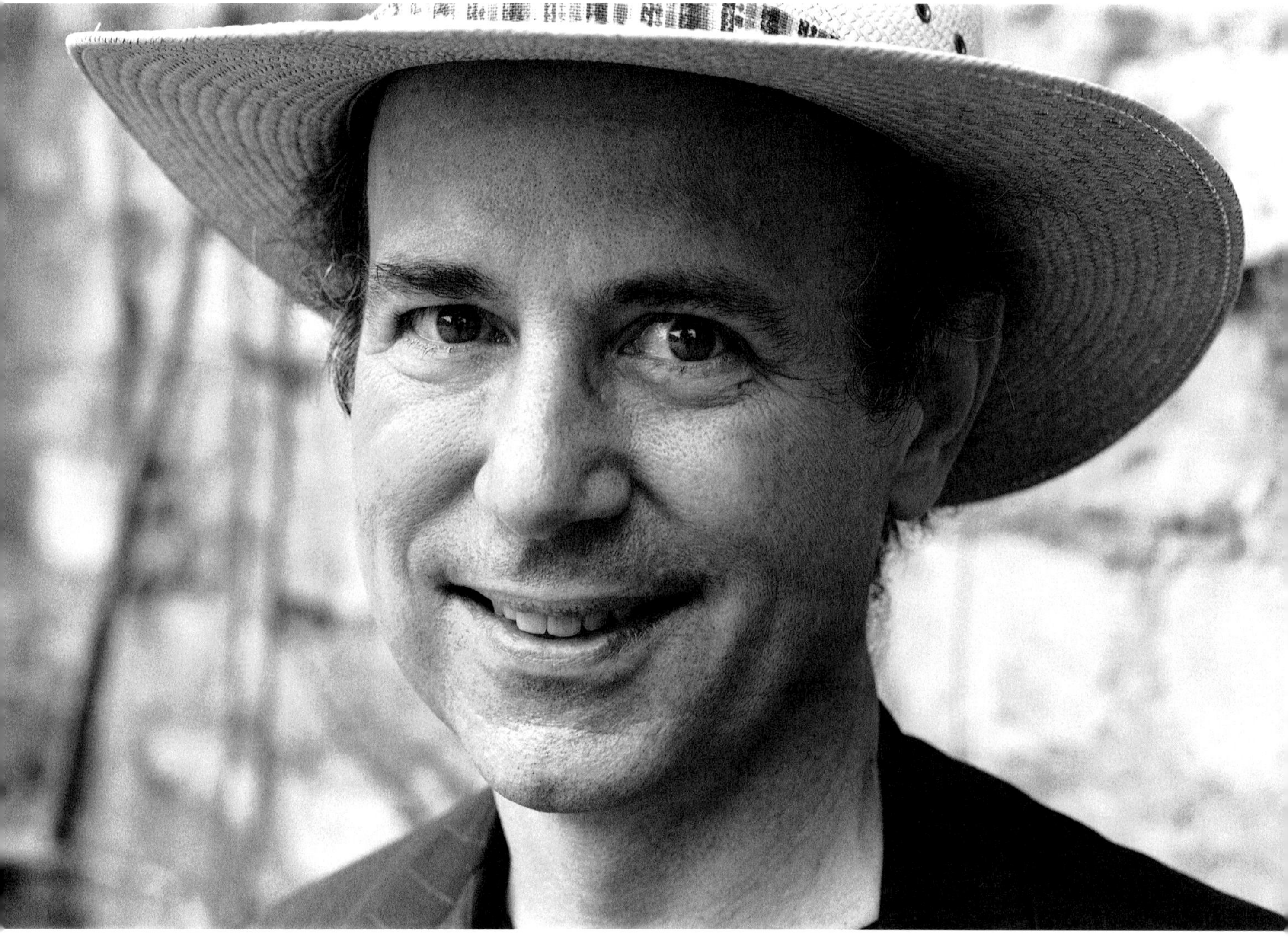

Kenneth G. Wilson *1936 †2013

The Nobel Prize in Physics 1982 was awarded to Kenneth G. Wilson "for his theory for critical phenomena in connection with phase transitions."

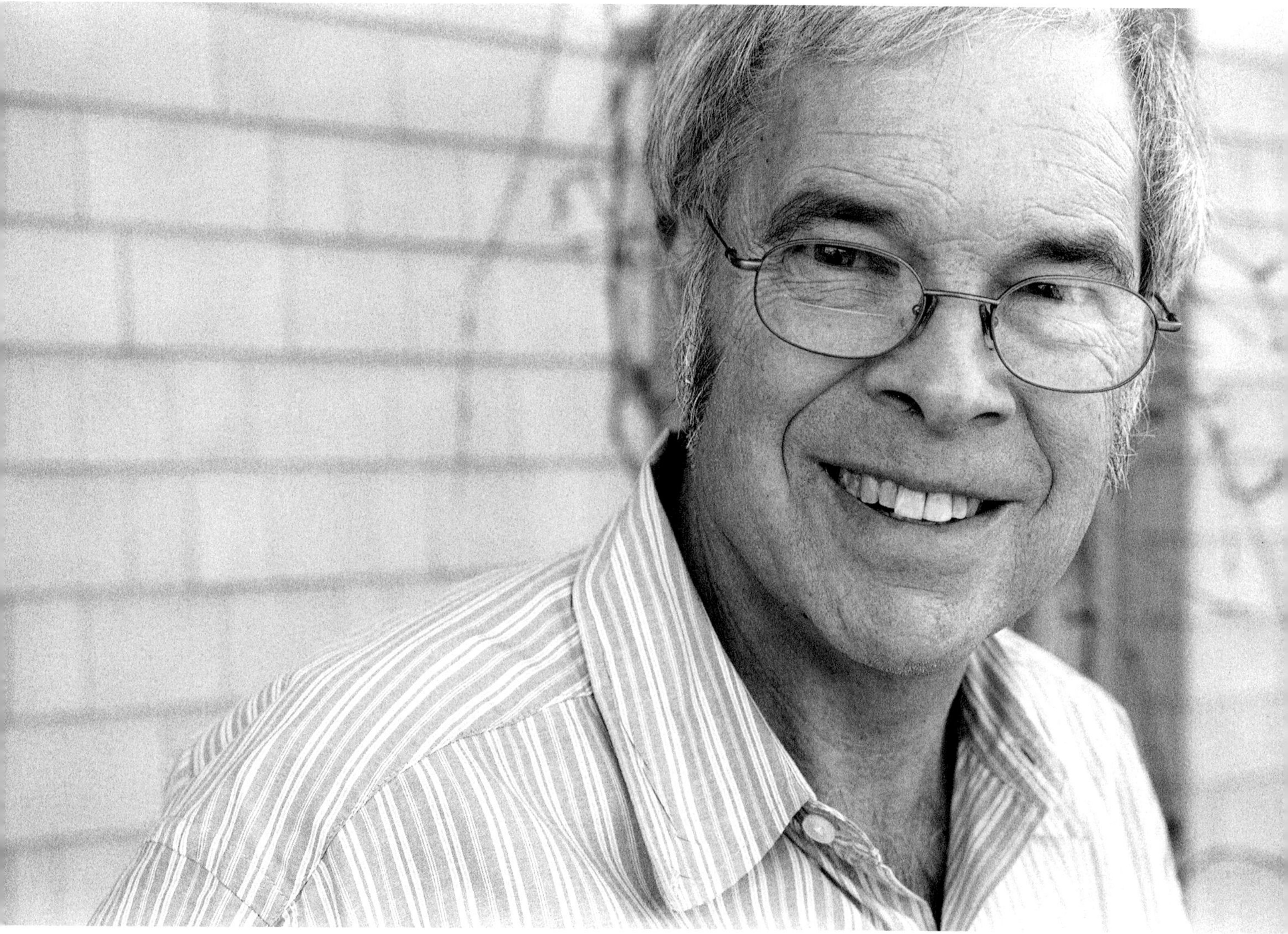

Robert Woodrow Wilson *1936

The Nobel Prize in Physics 1978 was divided, one half awarded to Pyotr Leonidovich Kapitsa "for his basic inventions and discoveries in the area of low-temperature physics", the other half jointly to Arno Allan Penzias and Robert Woodrow Wilson "for their discovery of cosmic microwave background radiation."

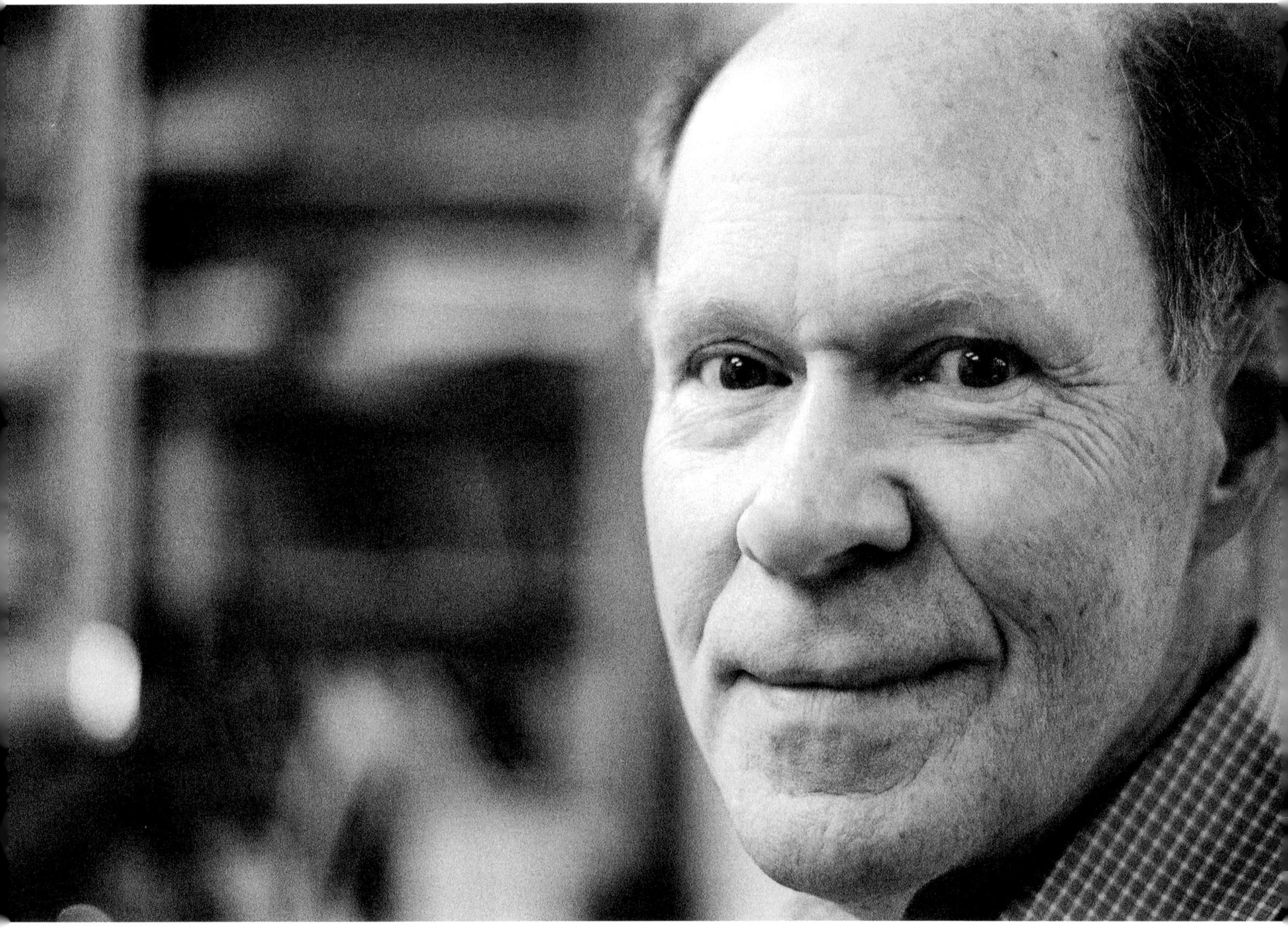

David J. Wineland *1944

The Nobel Prize in Physics 2012 was awarded jointly to Serge Haroche and David J. Wineland "for ground-breaking experimental methods that enable measuring and manipulation of individual quantum systems."

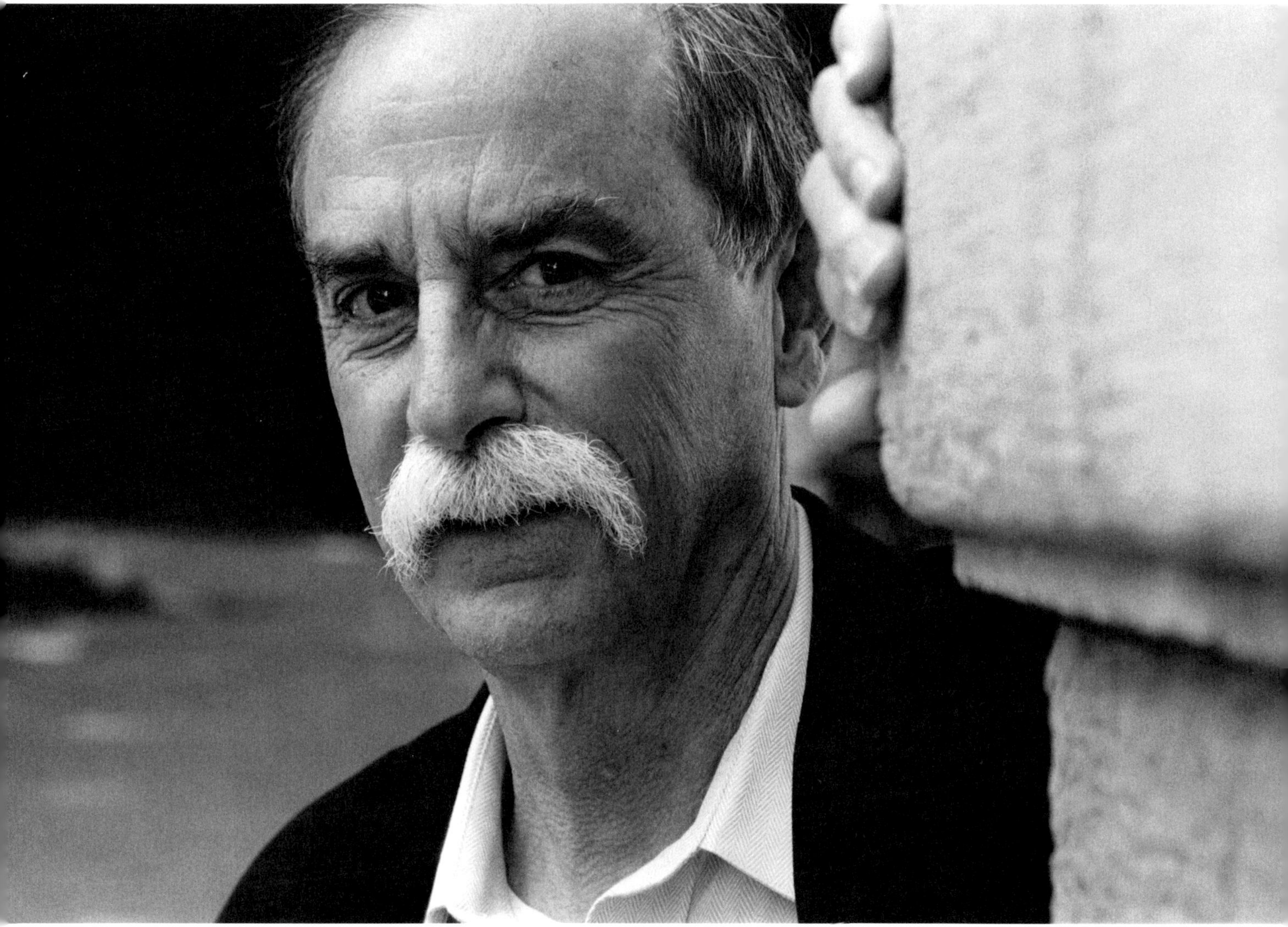

Chen Ning Yang *1922

The Nobel Prize in Physics 1957 was awarded jointly to Chen Ning Yang and Tsung-Dao (T. D.) Lee "for their penetrating investigation of the so-called parity laws which has led to important discoveries regarding the elementary particles."

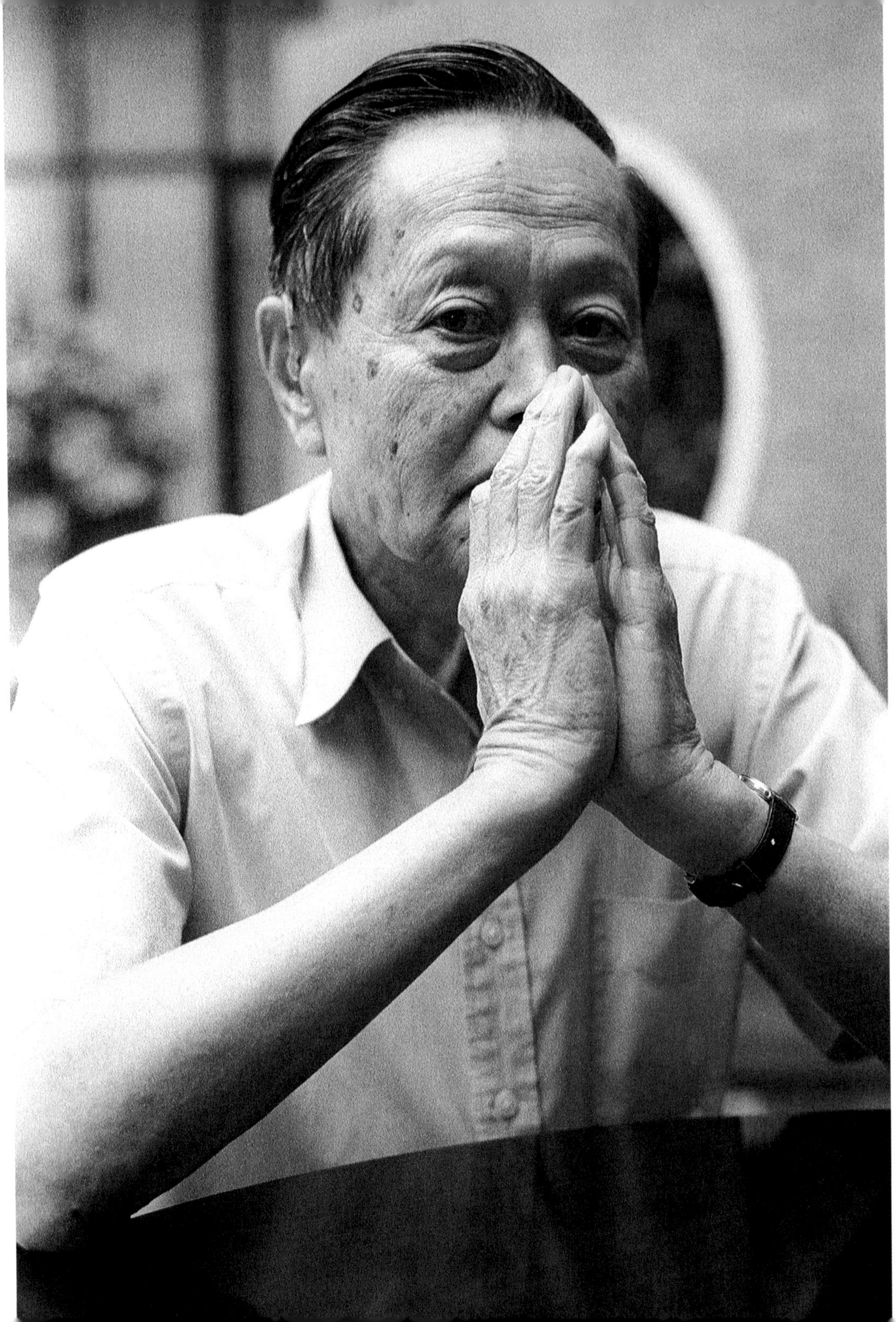

Photographic Index

Alexei A. Abrikosov, Stockholm, Sweden, 2003

Hiroshi Amano, Nagoya, Japan, 2015

Isamu Akasaki, Nagoya, Japan, 2015

Zhores I. Alferov, St. Petersburg, Russia, 2008

Philip Warren Anderson, Princeton, NJ, USA, 2003

Arthur Ashkin, Rumson, NJ, USA, 2018

Barry C. Barish, Los Angeles, CA, USA, 2017

J. Georg Bednorz, Rapperswil, Switzerland, 2016

Hans Albrecht Bethe, Ithaca, NY, USA, 2003

Gerd Binnig, Rüschlikon, Switzerland, 2000

Nicolaas Bloembergen, Lindau, Germany, 2000

Aage Niels Bohr, Copenhagen, Denmark, 2003

Willard S. Boyle, Stockholm, Sweden, 2009

Owen Chamberlain, Berkeley, CA, USA, 2004

Georges Charpak, Paris, France, 2003

Steven Chu, Lindau, Germany, 2000

Claude Cohen-Tannoudji, Lindau, Germany, 2015

Leon Neil Cooper, Lindau, Germany, 2003

Eric A. Cornell, Stockholm, Sweden, 2001

James Watson Cronin, Chicago, IL, USA, 2004

Raymond Davis Jr., Stockholm, Sweden, 2002

Hans G. Dehmelt, Seattle, WA, USA, 2004

François Englert, Brussels, Belgium, 2013

Leo Esaki, Kashiwa, Japan, 2013

Albert Fert, Paris, France, 2009

Val Logsdon Fitch, Princeton, NJ, USA, 2004

Jerome I. Friedman, Cambridge, MA, USA, 2004

Andre Geim, Manchester, UK, 2010

Murray Gell-Mann, New York, NY, USA, 2004

Pierre-Gilles de Gennes, Paris, France, 2000

Riccardo Giacconi, Washington, D.C., USA, 2007

Ivar Giaever, Mainau, Germany, 2010

Vitaly L. Ginzburg, Moscow, Russia, 2008

Donald Arthur Glaser, Lindau, Germany, 2000

Sheldon Lee Glashow, Berlin, Germany, 2007

Roy J. Glauber, Cambridge, MA, USA, 2007

David J. Gross, Lindau, Germany, 2010

Peter Grünberg, Jülich, Germany, 2007

F. Duncan M. Haldane, Princeton, NJ, USA, 2016

John L. Hall, Boulder, CO, USA, 2008

Theodor W. Hänsch, Munich, Germany, 2005

Serge Haroche, Lindau, Germany, 2013

Antony Hewish, Mainau, Germany, 2000

Peter W. Higgs, Edinburgh, UK, 2014

Gerardus 't Hooft, Lindau, Germany, 2000
Russell A. Hulse, Princeton, NJ, USA, 2003
Brian David Josephson, Lindau, Germany, 2016
Takaaki Kajita, Kashiwa, Japan, 2016
Charles Kuen Kao, Hong Kong, China 2010
Wolfgang Ketterle, Stockholm, Sweden ,2001
Jack S. Kilby, Stockholm, Sweden, 2000
Klaus von Klitzing, Lindau, Germany, 2015
Makoto Kobayashi, Tokyo, Japan, 2009
Masatoshi Koshiba, Lindau, Germany, 2004
J. Michael Kosterlitz, Providence, RI, USA, 2016
Herbert Kroemer, Santa Barbara, CA, USA, 2015
Willis Eugene Lamb, Lindau, Germany, 2004
Robert B. Laughlin, Lindau, Germany, 2001
Leon M. Lederman, Driggs, ID, USA, 2004
David M. Lee, Lindau, Germany, 2004
Tsung-Dao (T. D.) Lee, New York, NY, USA, 2002
Anthony J. Leggett, Stockholm, Sweden, 2003
Toshihide Maskawa, Kyoto, Japan, 2009
John C. Mather, Lindau, Germany, 2010
Arthur B. McDonald, Kingston, Ontario, Canada, 2015

Simon van der Meer, Geneva, Switzerland, 2004
Rudolf Ludwig Mössbauer, Lindau, Germany, 2000
Ben Roy Mottelson, Copenhagen, Denmark, 2004
Gérard Mourou, Paris, France, 2018
K. Alexander Müller, Zurich, Switzerland, 2000
Shuji Nakamura, Santa Barbara, CA, USA, 2015
Yoichiro Nambu, Chicago, IL, USA, 2009
Konstantin Novoselov, Manchester, UK, 2010
Douglas D. Osheroff, Lindau, Germany, 2008
Arno Allan Penzias, Lindau, Germany, 2010
Martin L. Perl, Stanford, CA, USA, 2003
Saul Perlmutter, Berkeley, CA, USA, 2012
William D. Phillips, Mainau, Germany, 2008
H. David Politzer, Pasadena, CA, USA, 2005
Norman F. Ramsey, Lindau, Germany, 2000
Robert C. Richardson, Lindau, Germany, 2000
Burton Richter, Stanford, CA, USA, 2004
Adam G. Riess, Stockholm, Sweden, 2011
Heinrich Rohrer, Madrid, Spain, 2000
Carlo Rubbia, Rome, Italy, 2002
Brian P. Schmidt, Canberra, Australia, 2011
John Robert Schrieffer, Tallahassee, FL, USA, 2005

Melvin Schwartz, Lindau, Germany, 2000

Kai M. Siegbahn, Angelholm, Sweden, 2003

George E. Smith, Stockholm, Sweden, 2009

George F. Smoot, Berkeley, CA, USA, 2006

Jack Steinberger, Lindau, Germany, 2008

Horst L. Störmer, Toulouse, France, 2000

Donna Strickland, Waterloo, Ontario, Canada, 2018

Joseph H. Taylor Jr., Princeton, NJ, USA, 2003

Richard E. Taylor, San Francisco, CA, USA, 2004

Kip Thorne, Oslo, Norway, 2016

David J. Thouless, Cambridge, UK, 2016

Samuel Chao Chung Ting, Lindau, Germany, 2000

Charles Hard Townes, Berkeley, CA, USA, 2008

Daniel C. Tsui, Lindau, Germany, 2000

Martinus J.G. Veltman, Lindau, Germany, 2000

Steven Weinberg, Austin, TX, USA, 2004

Rainer Weiss, Oslo, Norway, 2016

Carl E. Wieman, Stanford, CA, USA, 2016

Frank Wilczek, Lindau, Germany, 2005

Kenneth G. Wilson, Portland, ME, USA, 2003

David J. Wineland, Lindau, Germany, 2013

Robert Woodrow Wilson, Cambridge, MA, USA, 2003

Chen Ning Yang, Beijing, China, 2004

Educate – Inspire – Connect: Nobel Laureates as Role Models

The Lindau Nobel Laureate Meetings and Peter Badge's portraits are generating and promoting enthusiasm for science and research and the people behind it.

Since their foundation in 1951, the Lindau Nobel Laureate Meetings have developed into an internationally unique forum and venue for dialogue and scientific exchange. For nearly 70 years they have been globally recognised for the transfer of knowledge between laureates and young scientists and stand for the highest scientific standards as well as for international interactions.

Nobel Laureates and young scientists especially appreciate the large number of personal encounters and the one-of-a-kind atmosphere of the meetings. They encourage the intense exchange of ideas at the meetings. This dialogue between the different generations of scientists makes a significant contribution to the worldwide discourse, paving the way for interdisciplinary research and scientific progress. The inspiration by the Nobel Laureates, 'Rock Stars of Science' and big names that the young scientists usually only know from their textbooks, is often described by former participants as a once-in-a-lifetime experience. The Lindau Meetings are about people, not papers, and it is their unique format that serves as the birthplace of global networks of young scientists and doctoral students which will continue to make an impact far into the future.

The original idea of the meetings goes back to two Lindau physicians, Professor Dr. Gustav Parade and Dr. Franz Karl Hein. They intended to help the German scientific community in their own field of work to overcome the isolation they were experiencing as a result of the Nazi Regime and World War II. To this end, they wanted to establish a dialogue between young physicians and the leading international figures of their discipline. They found an enthusiastic supporter in

Count Lennart Bernadotte af Wisborg who very quickly became the real guiding spirit behind the Lindau Meetings. In 1951, the three initiated the first Lindau Nobel Laureate Meeting as a means of getting international meetings off the ground again.

The struggle for progress, reaching beyond existing limits, is a well-established tradition in the scientific community. This approach has also influenced and shaped the Lindau Meetings for almost 70 years. Here is where Nobel Laureates in physics, chemistry and physiology or medicine meet in alternate years for one week at Lake Constance. Since 2004 the laureates of the Sveriges Riksbank Prize in Economic Sciences in Memory of Alfred Nobel, who have also occasionally come to Lindau since 1970, have their own separate meetings every three years.

As in the first Lindau Meetings, the laureates today still appreciate the inimitable informal atmosphere that not only enables but also encourages an intensive exchange of ideas and memorable personal encounters. The fact that the Nobel Laureates are able to freely choose the subjects of their presentations has given the Lindau Meetings a special flavour. In addition, the benevolent support of the Nobel Foundation and the academic and scientific institutions awarding the Nobel Prizes has contributed significantly to the exceptional status of the Lindau Meetings.

Young people of different nationalities, cultures, disciplines and religions get together in Lindau. They discuss and deliberate scientific issues with the laureates enabling them to explore and understand both the differences and common ground, thus reinforcing the basis for peaceful cooperation. Defining mutual and universally true values is all the more important if significant ethical questions of modern science are to be evaluated and answered in a responsible manner.

The fall of the Iron Curtain has given Europe the chance to overcome the continent's internal division; the Lindau Meeting can now put into practice its vision of a »Window on the World«, especially in times when nationalism and isolationism worldwide is thriving. The dialogue with young scientists makes a valuable contribution to international communication and paves the way for discussion across disciplines, thus promoting scientific progress. Giving particularly talented and highly qualified young scientists the chance to meet the experts in their discipline is still as important as ever, even after nearly 70 years.

Indeed, even now in times of worldwide virtual networking, the Lindau Meetings with their unique, familiar flair have become not just more important but more attractive too. Conversation here provides the additional creative impulse which is the distinctive feature of the Lindau Nobel Laureate Meetings, as Roman Herzog, the former Federal President of Germany, has aptly described. There is hardly any other forum which offers quite the same opportunities. Dialogue and discussions taking place everywhere at the meeting, be it after the lectures, Agora Talks or panel discussions, in informal talks at Laureate Lunches, Science Walks or evening events, during the traditional boat trip from Lindau to Mainau Island on the last day of the meeting. This unique structure encourages the development of new ideas, through direct contacts, it inspires scientific work beyond just formulas and enables problems to be viewed critically from new angles. In short, it promotes new working approaches. The benefits of the meeting's structure can be seen far beyond the intensive use of the alumni dialogue. Many past participants have confirmed that a meeting in Lindau sows the seeds for long-lasting scientific and amicable bonds around the world.

Today the Council for the Lindau Nobel Laureate Meetings not only continues to pursue the original aims of the three founders, achieving these in a practical way again and again every year, but it also works to develop the meetings further.

Count Lennart Bernadotte acted as President of the Council for the Lindau Nobel Laureate Meetings for 38 years. This committee organises the meetings on a voluntary basis. His vision and charisma together with his roots in the Swedish Royal Family – his grandfather, later King Gustav V of Sweden, and his great grandfather, King Oscar II of Sweden, presented the first Nobel Prizes – and his connection to the Nobel institutions paved the way for the success of the Lindau Meetings. It was his vision that soon led to the original focus on medicine being extended to include the two other scientific Nobel disciplines. As a result, more than 30,000 students and young scientists from all continents have been able to visit the meetings since 1951, with the number of new alumni increasing by approximately 550 each year.

Even at the age of 95, the late Count Lennart Bernadotte embodied the »Spirit of Lindau«. He not only held the position of Honorary President of the Council and the Foundation, but is still held in great esteem since his death in December 2004.

In 1987 his wife, Countess Sonja Bernadotte, succeeded him as President of the Council. She has carried out this role just as dutifully and successfully. Under her auspices, the Lindau Meetings have managed to balance continuity and change the vision of the »Window on the World« has become a reality, not least because a network of partner institutions for nominating and selecting the young participants of the meetings had been established. After her early death in 2008, Countess Bettina Bernadotte succeeded her parents and has wholeheartedly continued and further developed the process of internationalisation and professionalisation. The Lindau Meetings have been part of her life since early childhood, thus, she is infused with the spirit of Lindau more than anybody else. To date, the Lindau Nobel Laureate Meetings interact closely with more than 200 of the most renowned research institutions worldwide to identify highly-talented young scientists.

At the occasion of the 50th anniversary of the Lindau Meetings in the summer of 2000, 50 participating Nobel Laureates, the Bernadotte family and the Council set up the Foundation Lindau Nobel Laureate Meetings under the chairmanship of Wolfgang Schürer, who served in this capacity for 15 years with great dedication and enormous success. Substantial donations to the foundations endowment from philanthropists, selected foundations, associations and international companies that are actively involved in supporting research, promoting science and intercultural dialogue, ensure the financial backing of the Lindau Meetings as well as the support by subsidies and contributions from the Federal Ministry of Education and Research, the Free State of Bavaria, scientific institutions etc.

Since the establishment in 2000, 340 laureates have joined the Founders' Assembly. Through their membership in this unique forum, Nobel Laureates demonstrate their strong support of the Lindau Meetings and entrust the Foundation with the consistent ongoing development of Lindau´s intergenerational dialogue.

Thanks to the initiative of Countess Sonja Bernadotte, the scientific secretaries of the Nobel Committees of the institutions which award the Nobel Prizes in Stockholm have been represented in the Council since the year 2000. This commitment ensures the highest scientific standards of quality for the meetings which are very flexible with regard to the planning of their programmes.

Mixing a familiar, informal meeting on the one hand and a programme rich in content and influenced by the laureates on the other, seems to be unique when compared to other international forums. Yet exactly this combination is the central feature of and secret behind the Lindau Meetings.

The »Spirit of Lindau« is characterised by excellence and its advancement. This is what the laureates, whose portraits were often taken by Peter Badge directly in Lindau, stand for. Peter Badge shows us the faces of impressive personalities whose scientific expertise provides the bedrock of their teaching. As well-respected representatives of the scientific community they stand for continual study and the addressing of universal issues. Their bright eyes convey a strength common to all of them and reveal their endeavours.

The concept of the Lindau Meetings is based on the methodology of the Nobel Laureates and relies on their personal commitment, for which we are eternally grateful. It allows, advances and accelerates learning through a dialogue between the generations. This is often seen as being the »brand« of the Lindau Nobel Laureate Meetings. Indeed, this is and always will be its intellectual basis, corresponding with the intention of the great Alfred Nobel, the award which bears his name and his laureates.

Nikolaus Turner
The author is a Member of the Board of the Council for the Lindau Nobel Laureate Meetings and a Member as well as Managing Director of the Executive Board of the Foundation Lindau Nobel Laureate Meetings (www.lindau-nobel.org).

Peter Badge

Photo: Jim Rakete

Peter Badge was born in Hamburg, Germany, in 1974 and began his career as a freelance artist and photographer in 1993. In 1995, he moved to Berlin to study art history and has been living and working there ever since. He initially worked as a freelance photographer for various magazines, and soon began developing his own independent concepts and projects. Choosing portraiture as his primary focus, Badge concentrates on noted personalities, including scientists and politicians, artists and actors, musicians and photographers and, in his latest works, on "Men on the Moon – from Armstrong to Aldrin", "Icons of the Economy" and "Philanthropists". He has also completed series on the electronic music pioneer Oskar Sala, Elvis impersonators ("Elviswho") and the popular German actor and musician Marius Müller-Westernhagen (in cooperation with the Art Cologne and the National Music Center in Washington, D.C.).

In 2000 Peter Badge embarked on a comprehensive series of photographs of Nobel Laureates. Commissioned by the Lindau Nobel Laureate Meetings in cooperation with the Smithsonian Institution, the National Portrait Gallery in Washington, D.C., the National Museum of American History as well as the Deutsche Museum and co-funded by the German foundation Klaus Tschira Stiftung. This project has developed into a long-term, ongoing mission, which has taken Badge all over the world in order to capture images of all living Nobel Laureates.

Peter Badge has published numerous books and has been printed in magazines and newspapers worldwide. His work is represented in various international private and public collections.